华罗庚–吴文俊数学出版基金资助项目

万哲先文选

万哲先／著

科学出版社

北京

内 容 简 介

万哲先院士著述甚丰,许多系统性的成果已总结在他的专著中. 除此以外,还有一些有意义的零散的研究工作. 本书挑选了其中的 9 篇论文,加上 10 部专著的序言和 6 篇回忆文章集结出版,以此庆祝万哲先院士九十华诞.

本书适合数学专业的研究人员、大学数学系的教师和学生,以及其他对数学有兴趣的年轻人阅读.

图书在版编目(CIP)数据

万哲先文选/万哲先著. —北京:科学出版社, 2017.6
ISBN 978-7-03-053025-7

Ⅰ. ①万⋯ Ⅱ. ①万⋯ Ⅲ. ①数学-文集 Ⅳ. ①O1-53

中国版本图书馆 CIP 数据核字 (2017) 第 116886 号

责任编辑:王丽平／责任校对:张凤琴
责任印制:张 伟／封面设计:黄华斌

科学出版社 出版
北京东黄城根北街 16 号
邮政编码:100717
http://www.sciencep.com

北京东华虎彩印刷有限公司 印刷
科学出版社发行 各地新华书店经销

*

2017 年 6 月第 一 版　开本:720×1000 1/16
2017 年 6 月第一次印刷　印张:14 1/4
字数:280 000
定价:89.00 元
(如有印装质量问题,我社负责调换)

目 录

A Proof for a Graphic Method for Solving the Transportation Problem ······ 1
Automorphisms and Isomorphisms of Linear Groups over Skew Fields ······ 9
Geometry of Classical Groups over Finite Fields and Its Applications ······ 15
Nonlinear Feedforward Sequences of m-sequences I ······ 40
A Linear Algebra Approach to Minimal Convolutional Encoders ········· 62
Representations of Forms by Forms in a Finite Field ······ 105
Geometry of Matrices ······ 136
On the Uniqueness of the Leech Lattice ······ 149
Symplectic Graphs and Their Automorphisms ······ 156
《典型群》序 ······ 173
《李代数》序 ······ 175
《有限几何与不完全区组设计的一些研究》序 ······ 177
《代数和编码》序 ······ 179
《非线性移位寄存器》序 ······ 183
Geometry of Classical Groups over Finite Fields: Preface ······ 184
Geometry of Matrices: Preface ······ 188
《有限典型群子空间的轨道生成的格》序 ······ 190
Lectures on Finite Fields and Galois Rings: Preface ······ 192
Finite Fields and Galois Rings: Preface ······ 194
怀念杨武之老师 ······ 196
回忆母校联大附中 ······ 199
Hsiao-Fu Tuan, 1914–2005 ······ 203
深深怀念陈省身先生 ······ 206
中国的代数学 ······ 209
忆华罗庚老师1950年回到清华园执教 ······ 219

A Proof for a Graphic Method for Solving the Transportation Problem*

Zhexian Wan (万哲先)**

(Institute of Mathematics, Academia Sinica)

1. The Method

In September, 1958, we learned a graphic method for solving the transportation problem from a certain transportation department in Peking. And we were asked to work out a mathematical proof for this method. The method may be stated as follows:

Suppose it is desired to set up a transportation schedule to distribute a certain goods from several sources to several destinations. The total amount of supply of all the sources is assumed to be equal to the total amount of demand of all the destinations. The schedule should be such that the total "transportation cost" (in terms of ton-kilometres, for instance) for realizing it would be a minimum.

Before giving the transportation schedule, we first draw a map showing all the sources and destinations, and their connecting lines (railways for instance), with the sources denoted by small circles "○" and the destinations by crosses "×". The numbers given beside the sources and those beside the destinations are their respective amounts (in terms of tons, say) of supply and demand.

---

* First reported in Chinese in *Shuxue Tongbao* (*Bulletin of Chinese Mathematics*), No. 11, pp. 478—482, 1958.

** In collaboration with Comrades Shen Xin-yao, Xu Yi-chao, Gan Dan-yan, Zhu Yong-jin, and Yang Xi-an.

Along the connecting lines are given the distances (in terms of kilometres, say) between the neighbouring cities or towns. Such a distribution map is shown in Fig. 1.

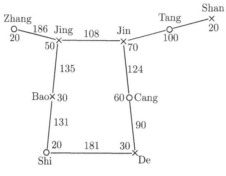

Fig. 1

Table 1 gives an illustrative schedule.

Table 1

Sources \ Destinations	Shan	Jin	Jing	De	Bao
Tang	0	70	30	0	0
Zhang	0	0	0	0	20
Cang	20	0	0	30	10
Shi	0	0	20	0	0

For realizing this schedule, a sort of flow diagram is required. Suppose there are goods, m tons say, flowing from source A to destination B. A flow vector is drawn alongside of the line segment representing the railway from A (on the right) to B (on the left) with the amount of flow m marked above the flow vector (see Fig. 2). If along the same line there are more than one flow vector in the same sense, a single flow vector is drawn with the total amount of transportation indicated, as illustrated in Fig. 3. If this is done for all the section lines, a complete flow diagram of the schedule can be drawn as in Fig. 4. It is evident that the total "transportation costs" (in terms of ton-kilometres, say) of schedules with the same flow diagram are equal.

A Proof for a Graphic Method for Solving the Transportation Problem

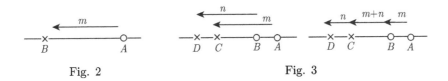

Fig. 2   Fig. 3

Clearly, when counter flow vectors occur in the flow diagram, for example, in the case of Fig. 4, when counter flows occur between the Jing-Bao and Jin-Tang lines, the transportation schedule is certainly not the best. In this case, the schedule should be changed so as to obtain a new one, of which the flow diagram contains no counter flows. For example, Fig. 4 may be changed into Fig. 5 in which counter flows disappear. The corresponding transportation schedule is shown in Table 2.

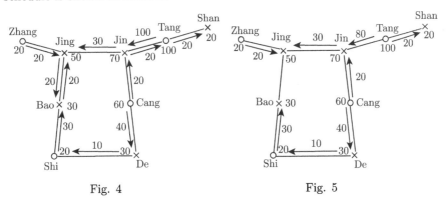

Fig. 4   Fig. 5

Table 2

Sources \ Destinations	Shan	Jin	Jing	De	Bao
Tang	20	70	10	0	0
Zhang	0	0	20	0	0
Cang	0	0	20	30	10
Shi	0	0	0	0	20

In the following we shall consider only schedules with flow diagrams that contain no counter flows.

Now let us consider a cycle C in a flow diagram; by a cycle is meant a closed path without self-intersecting points. There may be flow vectors in the positive sense of C (i.e. in the counterclockwise sense) and in the negative sense of C (i.e. in the clockwise sense). For distinction, flow vectors in the positive sense are placed outside the cycle, while those in the negative sense inside the cycle. Denote the total length of C by $S(C)$, that of the flow vectors in the positive sense of C by $S^+(C)$, and that of the flow vectors in the negative sense of C by $S^-(C)$. A cycle C is said to be normal if both $S^+(C) \leqslant \frac{1}{2}S(C)$ and $S^-(C) \leqslant \frac{1}{2}S(C)$ hold. A flow diagram is said to be normal, if each of its cycles is normal. Then the graphic method asserts: "A transportation schedule is an optimal one if and only if its flow diagram is normal"[1].

2. The Proof

Before we come to the proof of the above mentioned criterion, it may be remarked that branching lines of the distribution map can be omitted in the following manner. At a branching point is recorded only the net supply or the net demand, as the case may be, of the whole branching line (including the branching point), instead of the original amount of supply or demand. For instance, in Table 2, the Jin-Shan line can be omitted by changing Jin to a source with a net supply of 10 units and the Jing-Zhang line can also be omitted by considering Jing to be a destination with a net demand of 30 units. Thus we obtain the following distribution map that contains no branching lines (Fig. 6).

First let us consider the case where the flow diagram contains only a single cycle. For simplicity, we regard the demand of each destination to be 1 unit. In fact, if a destination demands m units, we may imagine a series of m destinations, each with a demand of 1 unit, situated at the original destination with distance 0 between any two such destinations.

---

[1] The original criterion, as we learned from the transportation company, states that a transportation schedule is optimal if and only if all those cycles which contain no other cycles inside themselves and all those which are not contained in any other cycle are normal. But this condition is not sufficient as may be shown by examples. The present criterion was suggested during the proof.

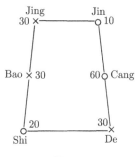

Fig. 6

Suppose that there is a transportation schedule with a non-normal flow diagram G (see Fig. 7). For definiteness, assume that $S^+(G) > \frac{1}{2}S(G)$. Then we may draw a new flow diagram G' which differs from G only in that the end point of each flow vector in the positive sense in G is now supplied by the flow vector in the negative sense in G', where other destinations are supplied by the same sources in G' as in G (see Fig. 8). (The process of change from G to G' is called the process of shrinking the flow vector in the positive sense by one station.) The "transportation cost" for transporting the goods to the end points of the flow vectors in the positive sense of G is equal to $S^+(G)$ according to G, and equal to $S^-(G')$ according to G'. Since $S^+(G) + S^-(G') \leqslant S(G)$ and $S^+(G) > \frac{1}{2}S(G)$, we have

$$S^-(G') < S^+(G).$$

Thus the transportation schedule with flow diagram G' is better than that with flow diagram G. This indicates that a schedule with a nonnormal map is not optimal.

Next, let us prove that transportation schedules with normal flow diagrams are always optimal. By the above proof, we know that normal flow diagrams always exist. If there is only one flow diagram, the statement just made is self-evident. Suppose now there are more than one flow diagram. Then we may arrange all the flow diagrams in an order $G_1, G_2, \cdots, G_n$, such that G_1 is the flow diagram without flow vectors in the negative sense, G_n is the flow diagram without flow vectors in the positive sense, and each G_{i+1} is obtained from the preceding G_i by the process of shrinking the flow vectors of G_i in the positive sense by one station ($i = 1, 2, \cdots, n-1$). Assume that G_i is the normal diagram with the smallest subscript and G_i is the normal diagram with the greatest subscript, then $i \leqslant j$, and the flow diagrams between G_i and G_j are all normal. If $i = j$, this is quite evident. Suppose now $i < j$. It is sufficient to show that any two neighbouring flow diagrams G_k and $G_{k+1} (i \leqslant k \leqslant j)$ have the same total "transportation cost". Notice that G_k and G_{k+1} differ only in the end points of the flow vectors in the positive sense of G_k. Since $S^+(G_k) \leqslant \frac{1}{2}S$, $S^-(G_{k+1}) \leqslant \frac{1}{2}S$, $S^-(G_{k+1}) \leqslant S - S^+(G_k)$, we have $S^+(G_k) = S^-(G_{k+1}) = \frac{1}{2}S$, S being the total length of the cycle. This proves that G_k and G_{k+1} have the same "transportation cost". Consequently, normal flow diagrams have the same "transportation cost". This finishes the proof for the case that the communication map consists of a single cycle.

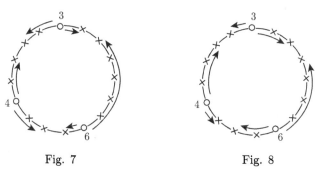

Fig. 7   Fig. 8

Now consider the general case. By the same argument as in the case of a single cycle, we can prove that transportation schedules with nonnormal flow diagrams are not optimal. It remains to show that schedules with normal flow diagrams have the same "transportation cost". Let G and G' be two distinct normal flow diagrams (Fig. 9), which, of course, belong to different schedules. In comparing G with G', we may assume that there are no common flow vectors in G and G'. In fact, if there is a common flow vector, we may remove a certain amount of transportation from both G and G' so as to obtain flow diagrams

G_1 and G'_1 without common flow vectors[2].

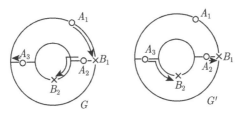

Fig. 9

Now we apply induction to the total amount of supply of all the sources of G. Let us start with a source A_1. Assume that, in G, one ton of goods is transported from A_1 to destination B_1. Then in G', one ton needed by B_1 is supplied by another source A_2. Then is G, A_2 supplies another destination B_3. Continuing in this way, and assuming that B_i is the first destination which coincides with a further destination $B_k (i < k)$, we obtain a closed path Z:

(*) $$B_i, A_{i+1}, B_{i+1}, \cdots, A_k, B_k = B_i.$$

In general, this is a self-intersecting closed path. Along this closed path, goods supplied according to G go round in one sense, while those supplied according to G' go round in another (see Fig. 9). It is evident that the closed path Z may be a single cycle or may be decomposed into several cycles $C_1, C_2, \cdots$. For example, the closed path shown in Fig. 10 may be decomposed into two cycles as shown in Fig. 11.

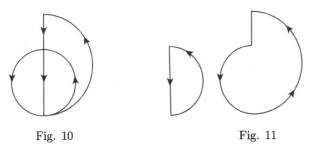

Fig. 10   Fig. 11

---

[2] This idea is due to Professor Tian Fang-zeng. We are indebted to Professor Tian for his valuable suggestions.

Then in each cycle, the flow vectors of G are all of the same sense, and those of G' are all of the same sense too. Since both G and G' are normal, the "transportation costs" for carrying one ton of goods according to G and G' in each of these cycles $C_i (i = 1, 2, \cdots)$ forming Z, are the same, i.e. both are equal to $\frac{1}{2}S(C_i)$. Thus the "transportation costs" for carrying one ton of goods according to G and G' in the closed path Z are also the same. Removing one ton of goods along the closed path Z from both G and G', we obtain two new normal flow diagrams G_1 and G'_1 with a smaller total amount of supply of all sources. By induction hypothesis, G_1 and G'_1 have the same amount of transportation, so do G and G'.

The proof is now complete.

Remarks 1. After we finished the above proof in September 1958, we learned that Professor Xu Guo-zhi and Comrade Gui Xiang-yun had also obtained a proof for the case of a communication map consisting of a single cycle by showing that the criterion of the graphic method is the same as that given by the simplex method. In 1959, the general case was also considered by them with the same method. But their proof has not yet been published.

2. That the transportation problem can be solved by the simplex method is well known. Recently, we learned that the problem had also been solved by Л. Б. Канторович and М. К. Гавурин by another method [1].

References

[1] Канторович, Л. В., Гавурин, М. К. Применение математических методов в вопросах анализа грузопотоков, *Сб. "проблемы повыщения эффективности работы транспорта"*, АН СССР. 1949, стр. 110–138.

Automorphisms and Isomorphisms of Linear Groups over Skew Fields

Hongshuo Ren, Zhexian Wan, and Xiaolong Wu

Let K be a skew field and $n \geq 2$ be an integer. Notations $\mathrm{GL}(n,K)$, $\mathrm{SL}(n,K)$, $\mathrm{PGL}(n,K)$ and $\mathrm{PSL}(n,K)$ are defined as usual. If $X \in \mathrm{SL}(n,K)$, we write $\overline{X}$ for its projection image in $\mathrm{PSL}(n,K)$. An automorphism Λ of $\mathrm{PSL}(n,K)$ is called standard if there exist a matrix $A \in \mathrm{GL}(n,K)$ and an automorphism σ or an antiautomorphism τ of K such that

$$\Lambda \overline{X} = \overline{AX^\sigma A^{-1}} \text{ for all } \overline{X} \in \mathrm{PSL}(n,K)$$

or

$$\Lambda \overline{X} = \overline{A(X'^\tau)^{-1} A^{-1}} \text{ for all } \overline{X} \in \mathrm{PSL}(n,K),$$

where X^σ (resp. X^τ) is the matrix obtained by the action of σ (resp. τ) on each entry of X, and X' is the transpose of X.

We have completed the proof of the following theorem [5, 6]:

Theorem *For any skew field K and any integer $n \geq 2$, all automorphisms of $\mathrm{PSL}(n,K)$ are standard.*

Remarks (1) The theorem can be generalized to any group Δ with $\mathrm{PSL}(n,K) \subseteq \Delta \subseteq \mathrm{PGL}(n,K)$.

(2) The theorem can be generalized to any group Δ with $\mathrm{SL}(n,K) \subseteq \Delta \subseteq$

1980 *Mathematics Subject Classification* (1985 *Revision*). Primary 20G35.

Projects supported by the Science Fund of the Chinese Academy of Sciences nos. 85188 and 85204.

$GL(n, K)$, where the definition of "standard" is modified by a homomorphism $\xi : \Delta \to$ the center of the multiplicative group of K.

(3) With minor modification of the proof of the theorem, we proved in [7] that if K and K_1 are skew fields, $n, n_1 \geqslant 2$ are integers and $\mathrm{PSL}(n, K)$ is isomorphic to $\mathrm{PSL}(n_1, K_1)$, then $n = n_1$ and K is isomorphic or anti-isomorphic to K_1, with the only exceptions: $\mathrm{PSL}(2, \mathbf{F}_4) \cong \mathrm{PSL}(2, \mathbf{F}_5)$ and $\mathrm{PSL}(2, \mathbf{F}_7) \cong \mathrm{PSL}(3, \mathbf{F}_2)$.

Proof of the Theorem O. Schreier and van der Waerden proved the theorem for commutative K in 1928 [8]. L. Hua corrected a mistake in their proof in 1948 [2]. J. Dieudonné proved the theorem for all $n \neq 2, 4$ in 1951 [1]. L. Hua solved the case $n = 4$ in 1951 [3]. L. Hua and Z. Wan solved the case $n = 2$, char $K > 0$ in 1953 [4]. Therefore, only the case $n = 2$, char $K = 0$ was left open.

Now write K^* for the multiplicative group of K, Z for the center of K, K^c for the commutator subgroup of K^*, and I for the identity matrix of order 2. Assume from now on that $n = 2$ and char $K = 0$. If S is a subset of $\mathrm{PSL}(2, K)$, denote by CS the centralizer of S in $\mathrm{PSL}(2, K)$ and by $C^2 S$ the set $\{X^2 | X \in CS\}$. If S is a group, denote by DS the commutator subgroup of S.

First assume $-1 \in K^c$. Write $\Lambda \begin{pmatrix} 1 & 0 \\ 0 & -1 \end{pmatrix} = \overline{A}$. Then we may assume $A = \begin{pmatrix} 0 & \alpha \\ 1 & 0 \end{pmatrix}$ for some $\alpha \in Z$. Write $\overline{B} = \Lambda^{-1} \begin{pmatrix} 1 & 0 \\ 0 & -1 \end{pmatrix}$. Then B is of the form $\begin{pmatrix} a & 0 \\ 0 & d \end{pmatrix}$ or $\begin{pmatrix} 0 & b \\ c & d \end{pmatrix}$. If $B = \begin{pmatrix} a & 0 \\ 0 & d \end{pmatrix}$ and $\overline{CC^2 A}$ contains an element $\begin{pmatrix} x & \alpha y \\ y & x \end{pmatrix}$, $xy \neq 0$, then $\begin{pmatrix} 1 & 0 \\ 0 & -1 \end{pmatrix} \notin \overline{CCC^2 A}$. But $\begin{pmatrix} a & 0 \\ 0 & d \end{pmatrix} \in \overline{CCC^2} \begin{pmatrix} 1 & 0 \\ 0 & -1 \end{pmatrix}$, so we get a contradiction. If $B = \begin{pmatrix} a & 0 \\ 0 & d \end{pmatrix}$ and $\overline{CC^2 A}$ contains no element of the form $\begin{pmatrix} x & \alpha y \\ y & x \end{pmatrix}$, $xy \neq 0$, then $\overline{CC^2 A}$ has at most four elements whereas $\overline{CC^2\begin{pmatrix} 1 & 0 \\ 0 & -1 \end{pmatrix}}$ has infinitely many elements.

So we again get a contradiction. Therefore, we must have $B = \begin{pmatrix} 0 & b \\ c & 0 \end{pmatrix}$.

Conjugating by $\begin{pmatrix} 1 & 0 \\ 0 & c^{-1} \end{pmatrix}$, we may assume $B = \begin{pmatrix} 0 & \beta \\ 1 & 0 \end{pmatrix}$, $\beta \in Z$. Hence

$$\begin{aligned} \text{PSL}(2, Z) &= DCC^2 \left\{ \overline{\begin{pmatrix} 1 & 0 \\ 0 & -1 \end{pmatrix}}, \overline{\begin{pmatrix} 0 & \alpha \\ 1 & 0 \end{pmatrix}} \right\} \\ &= DCC^2 \left\{ \overline{\begin{pmatrix} 1 & 0 \\ 0 & -1 \end{pmatrix}}, \overline{\begin{pmatrix} 0 & \beta \\ 1 & 0 \end{pmatrix}} \right\} \\ &= \text{PSL}(2, Z). \end{aligned}$$

By the known result for the commutative case, we may assume

$$\Lambda \overline{\begin{pmatrix} 1 & 1 \\ 0 & 1 \end{pmatrix}} = \overline{\begin{pmatrix} 1 & 1 \\ 0 & 1 \end{pmatrix}} \quad \text{and} \quad \Lambda \overline{\begin{pmatrix} 1 & 0 \\ 1 & 1 \end{pmatrix}} = \overline{\begin{pmatrix} 1 & 0 \\ 1 & 1 \end{pmatrix}}.$$

Now by the well-known method we can show that Λ is standard.

Now we assume $-1 \notin K^c$. The main steps of the proof are:

(1) Write $\Lambda \overline{\begin{pmatrix} 0 & -1 \\ 1 & 0 \end{pmatrix}} = \overline{S}$, $\Lambda \overline{\begin{pmatrix} 1 & 1/2 \\ 0 & 1 \end{pmatrix}} = \overline{W}$. The we may assume

$$S^2 = \alpha I, \quad (SW^2)^3 = \beta I, \quad (W^4 SWS)^3 = \gamma I, \quad \text{for some } \alpha, \beta, \gamma \in Z,$$

since $\begin{pmatrix} 0 & -1 \\ 1 & 0 \end{pmatrix}$ and $\begin{pmatrix} 1 & 1/2 \\ 0 & 1 \end{pmatrix}$ satisfy similar relations. A study of the elements of order three in PSL(2, K) shows $\beta, \gamma \in \{x^3 | x \in Z\}$. Hence by choosing representatives of S and W, we may assume $\beta = \gamma = 1$.

(2) We show that W is triangulizable. Define $M(2, K) = \left\{ \begin{pmatrix} a & b \\ c & d \end{pmatrix} \Big| a, b, c, d \in K \right\}$. We may regard $M(2, K)$ as an extension ring over K by the map $a \mapsto aI$, for all $a \in K$.

If $\alpha \notin K^2 = \{a^2 | a \in K\}$, we may assume $S = \begin{pmatrix} 0 & \alpha \\ 1 & 0 \end{pmatrix}$, $\alpha \in A$. Write $W = \begin{pmatrix} w & x \\ y & z \end{pmatrix}$. Define

$$A = \begin{pmatrix} y & -ywy^{-1} \\ 0 & 1 \end{pmatrix}, \quad T = AMA^{-1} = \begin{pmatrix} 0 & q \\ 1 & s \end{pmatrix} \text{ (say)}, \quad B = \begin{pmatrix} 0 & 1 \\ 1 & T \end{pmatrix}.$$

Then conjugate by BA in $\mathrm{GL}(2, M(2, K))$. We get

$$BASA^{-1}B^{-1} = \begin{pmatrix} * & * \\ R & * \end{pmatrix}, \quad \text{where } R = -\alpha y^{-1}((ywy^{-1} + yT)^2 - \alpha^{-1}I),$$

$$BAWA^{-1}B^{-1} = \begin{pmatrix} s - T & 1 \\ 0 & T \end{pmatrix}.$$

When R is invertible, we can conjugate by a matrix of the form $\begin{pmatrix} * & * \\ 0 & 1 \end{pmatrix}$ in $\mathrm{GL}(2, M(2, K))$ to send S and W to matrices of the form $\begin{pmatrix} 0 & \alpha \\ 1 & 0 \end{pmatrix}$ and $\begin{pmatrix} X & Y \\ 0 & T \end{pmatrix}$, respectively. If $X^2 + T^2$ is invertible, then we deduce from (1) that $(T^3 - I)(T^5 - I) = 0$. So T^{15} and hence W is triangulizable in $\mathrm{GL}(2, K)$. If $X^2 + T^2$ is not invertible, then neither is $T^4 - \alpha^{-1}I$ by (1). So T^4 and hence W is triangulizable. If R is not invertible, then we must have $R = 0$ since $\alpha \notin K^2$. Hence $(ywy^{-1} - yT)^2 - \alpha^{-1}I = 0$. But this means $WS = SW$, Which is impossible.

Now assume $\alpha \in K^2$. Then $\alpha = \theta^2$ for some $\theta \in Z$. We can assume $S = \begin{pmatrix} \theta & 0 \\ 0 & \theta \end{pmatrix}$. Write W, A, T, B as in the last paragraph. We have

$$BASA^{-1}B^{-1} = \begin{pmatrix} -\theta & 0 \\ R & \theta \end{pmatrix}, \quad \text{where } R = 2\theta(T - ywy^{-1}),$$

$$BAWA^{-1}B^{-1} = \begin{pmatrix} s - T & 1 \\ 0 & T \end{pmatrix}.$$

If R is invertible, we can proceed as in the case $\alpha \notin K^2$. If R is not invertible, then $T - ywy^{-1}$ is not invertible. So T and hence W is triangulizable in $GL(2,K)$.

(3) Write $S = \begin{pmatrix} 0 & 1 \\ \alpha & 0 \end{pmatrix}$, $\alpha \in Z$, $W = \begin{pmatrix} w & x \\ y & z \end{pmatrix}$. If $y = 0$, we can use equalities in (1) to get $w = z^{-1} \in Z$ and $x = (w + w^{-1})^{-1}$. So

$$\mathrm{PSL}(2,Z) = DC\left\{\overline{\begin{pmatrix} 0 & -1 \\ 1 & 0 \end{pmatrix}}, \overline{\begin{pmatrix} 1 & 1/2 \\ 0 & 1 \end{pmatrix}}\right\} = DC\{\overline{S}, \overline{W}\} = \mathrm{PSL}(2,Z)$$

and we are through. If $x = 0$, we can conjugate by $\begin{pmatrix} 0 & 1 \\ \alpha & 0 \end{pmatrix}$ and return to the case $y = 0$. If $xy \neq 0$, then we must have $uxu + zu - uw - y = 0$ for some $u \in K$ since W is triangulizable. When $u^2 \neq \alpha$, we can conjugate by $\begin{pmatrix} 1 & \alpha^{-1}u \\ u & 1 \end{pmatrix}$ and return to the case $y = 0$, When $u^2 = \alpha$, we can conjugate by $\begin{pmatrix} 1 & 0 \\ u & 0 \end{pmatrix}$ and then use equalities in (1) to get a contradiction. This completes the proof.

References

[1] J. Dieudonné, *On the automorphisms of classical groups*, Mem. Amer. Math. Soc. No. 2 (1951).

[2] L. Hua, *On the automorphisms of the symplectic groups over any field*, Ann. of Math. (2) **49** (1948), 739–759.

[3] L. Hua, *Supplement to the paper of Dieudonné on the automorphisms of classical groups*, Mem. Amer. Math. Soc. No. 2 (1951), 96–122.

[4] L. Hua and Z. Wan, *Automorphisms ans isomorphisms of linear groups*, Acta. Math. Sincia **2** (1953), 1–32. (Chinese)

[5] H. Ren and Z. Wan, *Automorphisms of $\mathrm{PSL}_2^{\pm}(K)$ over any skew field K*, Acta Math. Sinica **25** (4)(1982), 484–492.

[6] H. Ren, Z. Wan, and X. Wu, *Automorphisms of $\mathrm{PSL}(2,K)$ over skew fields*, Acta Math. Sinica (to appear).

[7] H. Ren, Z. Wan, and X. Wu, *Isomorphisms of linear groups over skew fields*, (to appear).

[8] O. Schreier and van der Waerden, *Die Automorphismen der projektive Gruppen*, Abh. Math. Sem. Univ. Hamburg **6** (1928), 303–322.

Institute of Mathematics, Chinese Academy of Sciences

Institute of System Sciences, Chinese Academy of Sciences

Institute of Mathematics, Chinese Academy of Sciences

Geometry of Classical Groups over Finite Fields and Its Applications

Zhexian Wan[a,b]

[a] Institute of Systems Science, Chinese Academy of Sciences, Beijing, 100080, China
[b] Department of Information Technology, Lund University, Lund, Sweden

Received 15 November 1994

Abstract This is a survey paper on the geometry of classical groups over finite fields and its applications, old and new. The contents of the paper are as follows:

(1) The seed of our study; (2) The problems we are interested in; (3) History; (4) Recent results; (5) Application to association schemes and designs; (6) Application to authentication codes; (7) Application to projective codes; (8) Application to lattices generated by orbits of subspaces; (9) Application to representations of forms by forms; (10) Concluding remarks.

1. The Seed of Our Study

Let $\mathbb{F}_q$ be a finite field with q elements, where q is a prime power, $\mathbb{F}_q^{(n)}$ be the n-dimensional row vector space over $\mathbb{F}_q$, and $\mathrm{GL}_n(\mathbb{F}_q)$ be the *general linear group* of degree n over $\mathbb{F}_q$. $\mathrm{GL}_n(\mathbb{F}_q)$ acts on $\mathbb{F}_q^{(n)}$ in the following way:

$$\begin{aligned} \mathbb{F}_q^{(n)} \times \mathrm{GL}_n(\mathbb{F}_q) &\to \mathbb{F}_q^{(n)}, \\ ((x_1, x_2, \cdots, x_n), T) &\mapsto (x_1, x_2, \cdots, x_n)T. \end{aligned} \qquad (1)$$

Let P be an m-dimensional subspace of $\mathbb{F}_q^{(n)}$ and $v_1, v_2, \cdots, v_m$ be a basis of P, then

$$\begin{pmatrix} v_1 \\ v_2 \\ \vdots \\ v_m \end{pmatrix} \qquad (2)$$

is an $m \times n$ matrix of rank m over $\mathbb{F}_q$. We call the matrix (2) a *matrix representation* of the subspace P and use also the same letter P to denote the matrix (2) if no ambiguity arises. The action (1) of $\mathrm{GL}_n(\mathbb{F}_q)$ on $\mathbb{F}_q^{(n)}$ induces an action on the set of subspaces of $\mathbb{F}_q^{(n)}$ such that $T \in \mathrm{GL}_n(\mathbb{F}_q)$ carries the subspace P to PT. We may propose the following problems:

(i) What are the orbits of subspaces of $\mathbb{F}_q^{(n)}$ under the action of $\mathrm{GL}_n(\mathbb{F}_q)$?

(ii) How many orbits are there?

(iii) What are the lengths of the orbits?

(iv) What is the number of subspaces in an orbit contained in a given subspace?

The answers to these four problems are well known, they are:

(i) Two subspaces belong to the same orbit if and only if their dimensions are equal.

(ii) There are altogether $n+1$ orbits.

(iii) Denote the length of the orbit of m-dimensional subspaces ($0 \leqslant m \leqslant n$) by $N(m,n)$, then

$$N(m,n) = \frac{\prod_{i=n-m+1}^{n}(q^i-1)}{\prod_{i=1}^{m}(q^i-1)}. \qquad (3)$$

(iv) The number of k-dimensional subspaces contained in a given m-dimensional subspace ($0 \leqslant k \leqslant m \leqslant n$) is $N(k,m)$.

2. The Problems We Are Interested in

It is natural to propose the following problems:

Use any one of the other classical groups, such as the symptectic group $\mathrm{Sp}_n(\mathbb{F}_q)$ (where $n = 2v$), the unitary group $U_n(\mathbb{F}_q)$ (where q is a square) and the orthogonal group $O_n(\mathbb{F}_q)$ (where $n = 2v + \delta$ and $\delta = 0, 1$, or 2) to replace $\mathrm{GL}_n(\mathbb{F}_q)$, then study problems (i)–(iv).

The above problems look quite classical. For instance, they could have been proposed in 1900, the year Dickson published his famous monograph *Linear Groups* [16], in which all the ingredients for proposing them are contained. But as far as I know, mathematicians started to study these problems much later.

3. The History of the Problems

In 1937 Witt [72] studied the above mentioned problem (i) for the orthogonal group over any field F of characteristic $\neq 2$. Let S be an $n \times n$ nonsingular symmetric matrix over F. The *orthogonal group* of degree n over F relative to S, denoted by $O_n(F, S)$, is defined to be

$$O_n(F, S) = \{T \in \mathrm{GL}_n(F) | TS^tT = S\}. \tag{4}$$

The famous Witt's theorem asserts that two subspaces P_1 and P_2 of $F^{(n)}$ belong to the same orbit of $O_n(F, S)$ if and only if $\dim P_1 = \dim P_2$ and $P_1 S^t P_1$ and $P_2 S^t P_2$ are cogredient. Later Witt's theorem was generalized to other classical groups by Arf [1], Dieudonné [17, 18], Hua [26], etc. [14, 15, 38]. It is worth to mention that Hua [26] gave also a short and elegant matrix proof of the generalized Witt's theorem, which was reproduced in [59].

In his invited address at the International Congress of Mathematicians, Edinburgh, 1958, Segre [43] announced his formula for the number of $(m-1)$-dimensional flats lying on a nondegenerate quadric in $\mathrm{PG}\,(n-1, \mathbb{F}_q)$, the $(n-1)$-dimensional projective space over $\mathbb{F}_q$:

$$\frac{\prod_{i=v-m+1}^{v}(q^i - 1)(q^{i+\delta-1} - 1)}{\prod_{i=1}^{m}(q^i - 1)}, \tag{5}$$

where $1 \leqslant m \leqslant v$, $v = (n-1)/2$ and $\delta = 1$ when n is odd, and $v = n/2$, $\delta = 0$ or $v = n/2 - 1$, $\delta = 2$ when n is even and the quadric is of the hyperbolic type or the elliptic type, respectively. The complete proof of this formula appeared one year later [44]. He used a geometric method to deduce this formula, which works for both the case of characteristic not two and the case of characteristic two. In 1962 Ray-Chaudhuri [41] also deduced Segre's formula (5).

For simplicity, assume that q is odd. Then the $(m-1)$-dimensional flats lying on a nondegenerate quadric in PG $(n-1, \mathbb{F}_q)$ are m-dimensional totally isotropic subspaces of $\mathbb{F}_q^{(n)}$ with respect to the $n \times n$ nonsingular symmetric matrix S defining the quadric, which, by Witt's theorem, form an orbit of subspaces under the orthogonal group $O_n(\mathbb{F}_q, S)$. Thus, Segre actually studied the above-mentioned problem (iii) for the orthogonal group over $\mathbb{F}_q$, but he restricted himself to consider only the orbits of totally isotropic or totally singular subspaces corresponding to cases when q is odd or even, respectively.

In 1964 Dai, Feng, Yang, three students of mine at that time and myself [15, 20, 47, 69] studied problem (iii) for the groups $Sp_{2v}(\mathbb{F}_q)$, $U_n(\mathbb{F}_q)$ (where q is a square), and $O_{2v+\delta}(\mathbb{F}_q)$ (where $\delta = 0, 1$, or 2). We determined not only the lengths of those orbits of totally isotropic or totally singular subspaces but also the lengths of all the orbits. We call these results the Anzahl theorems in the geometries of these classical groups. Our methods were algebraic and our results were compiled in our monograph [67]. In 1965 Pless [39] computed the lengths of the orbits of totally isotropic subspaces of $\mathbb{F}_q^{(2v)}$ under the group $Sp_{2v}(\mathbb{F}_q)$ and the number of totally isotropic subspaces of the same dimension of $\mathbb{F}_q^{(2v+\delta)}$ (where $\delta = 1$ or 2) with respect to a $(2v+\delta) \times (2v+\delta)$ nonsingular non-alternate symmetric matrix over $\mathbb{F}_q$ when q is even. In 1965 Segre [45] and in 1966 Bose and Chakravarti [3] determined the lengths of the orbits of totally isotropic subspaces of $\mathbb{F}_q^{(n)}$ under the group $U_n(\mathbb{F}_q)$ (where q is a square).

In 1966 the author studied problem (iv) for the group $Sp_{2v}(\mathbb{F}_q)$, $U_n(\mathbb{F}_q)$ (q is a square), and $O_{2v+\delta}(\mathbb{F}_q)$ (where $\delta = 0, 1$, or 2) and obtained closed formulas for the number of subspaces in an orbit under each of these groups contained in a given subspace. These results are also called Anzahl theorems and were compiled in [67] too.

4. Recent Results

In the early 1990s I returned to the study of the geometry of classical groups over finite fields and obtained the following results:

(1) Problems (i) and (ii) for the symplectic, unitary, and orthogonal groups over finite fields are studied [51, 54, 60, 61]. Of course, Witt's theorem and its generalizations give a solution to problem (i), but we would like to use a set of numerical invariants to characterize an orbit and to derive the conditions satisfied by them when such an orbit exists, then the number of orbits can be computed.

Take the symplectic case as an example. Let

$$K = \begin{pmatrix} 0 & I^{(v)} \\ -I^{(v)} & 0 \end{pmatrix}. \tag{6}$$

Then the symplectic group of degree $2v$ is defined as

$$\mathrm{Sp}_{2v}(\mathbb{F}_q) = \{T \in \mathrm{GL}_{2v}(\mathbb{F}_q) | TK^tT = K\}. \tag{7}$$

Let P be an m-dimensional subspace of $\mathbb{F}_q^{(2v)}$. Clearly, PK^tP is alternate and, hence, the rank of PK^tP is even. Assume that rank of $PK^tP = 2s$, then P is said to be of type(m,s). From Dieudonné's generalization [17] of Witt's theorem it follows that two subspaces belong to the same orbit under $\mathrm{Sp}_{2v}(\mathbb{F}_q)$ if and only if they are of the same type. It can be proved [51] that the type(m,s) of a subspace satisfies the inequality

$$2s \leqslant m \leqslant v + s \tag{8}$$

and that for any pair of nonnegative integers (m, s) satisfying (8) there exist subspaces of type (m, s). Thus, the number of orbits of subspaces under $\mathrm{Sp}_{2v}(\mathbb{F}_q)$ is equal to the number of pairs of nonnegative integers (m, s) satisfying (8). We computed that the latter is equal to

$$\frac{1}{2}(v+1)(v+2). \tag{9}$$

By the way we mention that the length $N(m, s; 2v)$ of the orbit of subspaces of type (m, s) of $\mathbb{F}_q^{(2v)}$ given in [47] is

$$N(m, s; 2v) = q^{2s(v+s-m)} \frac{\prod_{i=v+s-m+1}^{v}(q^{2i}-1)}{\prod_{i=1}^{s}(q^{2i}-1)} \prod_{i=1}^{m-2s}(q^i-1). \tag{10}$$

This is the solution to problem (iii) for the symplectic group.

(2) The singular symplectic, unitary, and orthogonal groups are introduced and the problems (i)–(iv) are studied [62,63].

Take the singular symplectic case as an example. Let

$$K_l = \begin{pmatrix} K & \\ & 0^{(l)} \end{pmatrix}, \tag{11}$$

where K is the nonsingular alternate matrix (6). Define

$$\mathrm{Sp}_{2v+l,v}(\mathbb{F}_q) = \{T \in \mathrm{GL}_{2v+l}(\mathbb{F}_q) | TK_l^\mathrm{t}T = K_l\}, \tag{12}$$

which is called the *singular symplectic group* over $\mathbb{F}_q$. Clearly, $\mathrm{Sp}_{2v+l,v}(\mathbb{F}_q)$ acts on $\mathbb{F}_q^{(2v+l)}$ in an obvious way. Then problems (i)–(iv) can be studied for $\mathrm{Sp}_{2v+l,v}(\mathbb{F}_q)$, and complete results are obtained.

Similarly, singular unitary and orthogonal groups over $\mathbb{F}_q$ can be defined, and complete results for problems (i)–(iv) are obtained.

A natural question arises. Why do we study the geometry of singular symplectic, unitary, and orthogonal groups over finite fields?

The answer to problem (iv) for the general linear group $\mathrm{GL}_n(\mathbb{F}_q)$ is easy: the number of k-dimensional subspaces contained in a given m-dimensional subspace ($0 \leqslant k \leqslant m \leqslant n$) of $\mathbb{F}_q^{(n)}$ is $N(k, m)$. However, problem (iv) for the other classical groups is not so easy.

Take again the symplectic case as an example. Now assume that $\mathrm{Sp}_{2v}(\mathbb{F}_q)$ acts on $\mathbb{F}_q^{(2v)}$. Given a subspace P of type (m, s), where (m, s) satisfies (8), we would like to compute the number of subspaces of type (m_1, s_1), where $2s_1 \leqslant m_1 \leqslant v+s_1$, contained in P. Denote this number by $N(m_1, s_1; m, s; 2v)$. We may choose a matrix representation of P, denoted by P again, such that

$$PK^tP = \begin{pmatrix} 0 & I^{(s)} & \\ -I^{(s)} & 0 & \\ & & 0^{(m-2s)} \end{pmatrix}. \tag{13}$$

Let P_1 be a subspace of type (m_1, s_1) contained in P. As an m_1-dimensional subspace of the m-dimensional space P, P_1 has a matrix representation, denoted by P_1 again, which is an $m_1 \times m$ matrix of rank m_1. Then as a subspace of $\mathbb{F}_q^{(2v)}$, the subspace P_1 has $P_1 P$ as a matrix representation. Similarly, we can choose the matrix P_1 such that

$$(P_1 P) K^t (P_1 P) = \begin{pmatrix} 0 & I^{(s_1)} & \\ -I^{(s_1)} & 0 & \\ & & 0^{(m_1 - 2s_1)} \end{pmatrix}. \tag{14}$$

Then

$$P_1 \begin{pmatrix} 0 & I^{(s)} & \\ -I^{(s)} & 0 & \\ & & 0^{(m-2s)} \end{pmatrix} {}^t P_1 = \begin{pmatrix} 0 & I^{(s_1)} & \\ -I^{(s_1)} & 0 & \\ & & 0^{(m_1 - 2s_1)} \end{pmatrix}. \tag{15}$$

Thus, for any $T \in \mathrm{Sp}_{2s+(m-2s),s}(\mathbb{F}_q)$, $P_1 T P$ is also a matrix representation of a subspace of type (m_1, s_1) and contained in P and as a subspace of P it is represented by the matrix $P_1 T$. Therefore, it is natural to introduce the singular symplectic group $\mathrm{Sp}_{2s+(m-2s),s}(\mathbb{F}_q)$ and study how the subspaces of $\mathbb{F}_q^{(m)}$ are subdivided into orbits under $\mathrm{Sp}_{2s+(m-2s),s}(\mathbb{F}_q)$, the length of each orbit, and what orbits are contained in P.

(3) For pseudo-symplectic groups over finite fields of characteristic 2 problems (i)–(iv) are also studied [36, 56].

Now let $\mathbb{F}_q$ be a finite field of characteristic 2, than any $n \times n$ nonsingular non-alternate symmetric matrix over $\mathbb{F}_q$ is cogredient to either

$$S_1 = \begin{pmatrix} 0 & I^{(v)} & \\ I^{(v)} & 0 & \\ & & 1 \end{pmatrix} \quad \text{when } n = 2v+1 \text{ is odd} \tag{16}$$

or

$$S_2 = \begin{pmatrix} 0 & I^{(v)} & & \\ I^{(v)} & 0 & & \\ & & 0 & 1 \\ & & 1 & 1 \end{pmatrix} \quad \text{when } n = 2v+2 \text{ is even.} \tag{17}$$

We use S_δ ($\delta = 1$ or 2) to cover these two cases. Define the *pseudo-symplectic group* of degree $2v + \delta$ over $\mathbb{F}_q$ to be

$$\mathrm{Ps}_{2v+\delta}(\mathbb{F}_q) = \{T \in \mathrm{GL}_{2v+\delta}(\mathbb{F}_q) | TS_\delta{}^t T = S_\delta\}. \tag{18}$$

It was proved by Dieudonné [17] that $\mathrm{Ps}_{2v+1}(\mathbb{F}_q) \simeq \mathrm{Sp}_{2v}(\mathbb{F}_q)$ and $\mathrm{Ps}_{2v+2}(\mathbb{F}_q)$ has a normal series with $\mathrm{Sp}_{2v}(\mathbb{F}_q)$ as one of its factors and with the additive group of $\mathbb{F}_q$ as all the other factors. Thus, from a group theory point of view the pseudo-symplectic group $\mathrm{Ps}_{2v+\delta}(\mathbb{F}_q)$ is less interesting. However, its geometry is very peculiar. Let P be an m-dimensional subspace of $\mathbb{F}_q^{(2v+\delta)}$, then $PS_\delta{}^t P$ is a symmetric matrix and is cogredient to one of the following normal forms:

$$\begin{pmatrix} 0 & I^{(s)} & \\ I^{(s)} & 0 & \\ & & 0^{(m-2s)} \end{pmatrix}, \tag{19}$$

$$\begin{pmatrix} 0 & I^{(s)} & & \\ I^{(s)} & 0 & & \\ & & 1 & \\ & & & 0^{(m-2s-1)} \end{pmatrix}, \tag{20}$$

and

$$\begin{pmatrix} 0 & I^{(s)} & & & \\ I^{(s)} & 0 & & & \\ & & 0 & 1 & \\ & & 1 & 1 & \\ & & & & 0^{(m-2s-2)} \end{pmatrix}. \tag{21}$$

P is called a subspace of type $(m, 2s+\tau, s, \varepsilon)$, where $\tau = 0, 1$, or 2 corresponding to the above three normal forms (19), (20), or (21), respectively, $\varepsilon = 0$ or 1 corresponding to the cases $e_{2v+1} \notin P$ or $e_{2v+1} \in P$, respectively, and e_{2v+1} is the $(2v+\delta)$-dimensional row vector whose $(2v+1)$th component is 1 and other components are all 0's. It is proved [36] that two subspaces of $\mathbb{F}_q^{(2v+\delta)}$ belong to the same orbit under $\mathrm{Ps}_{2v+\delta}(\mathbb{F}_q)$ if and only if they are of the same type and that subspaces of type $(m, 2s+\tau, s, \varepsilon)$ exist if and only if

$$(\tau, \varepsilon) = \begin{cases} (0,0), (1,0), (1,1), \text{ or } (2,0), & \text{when } \delta = 1, \\ (0,0), (0,1), (1,0), (2,0), \text{ or } (2,1), & \text{when } \delta = 2 \end{cases} \quad (22)$$

and

$$2s + \max\{\tau, \varepsilon\} \leqslant m \leqslant v + s + [(\tau + \delta - 1)/2] + \varepsilon. \quad (23)$$

Using conditions (22) and (23) we can compute the number of orbits of subspaces under $\mathrm{Ps}_{2v+\delta}(\mathbb{F}_q)$, which is equal to

$$\frac{1}{2}(v+1)((v+4)\delta + 3v). \quad (24)$$

Denote the length of the orbit of subspaces of type $(m, 2s+\tau, s, \varepsilon)$ by $N(m, 2s+\tau, s, \varepsilon; 2v+\delta)$. Then

$$N(m, 2s+\tau, s, \varepsilon; 2v+\delta) = q^{n_0 + 2(s+(2-\delta)[\tau/2])(v+s!-m+\delta[(\tau+1)/2]+(\delta!-1)(\tau-1)(\tau!-2)\varepsilon/2)}$$

$$\times \frac{\prod_{i=v+s-m+[(\tau+\delta-1)/2]+\varepsilon+1}^{v}(q^{2i}-1)}{\prod_{i=1}^{s}(q^{2i}-1) \prod_{i=1}^{m-2s-\max(\tau,\varepsilon)}(q^i-1)}, \quad (25)$$

where $n_0 = 1$ when $\delta = 1$, and $n_0 = m, 0, 2(v+1) - m, 2(v+1) - m$, or $2(v+1) - m$ corresponding to the cases $(\tau, \varepsilon) = (0,0), (0,1), (1,0), (2,0)$, or $(2,1)$, respectively, when $\delta = 2$.

In order to study problem (iv) for the pseudo-symplectic group $\mathrm{Ps}_{2v+\delta}(\mathbb{F}_q)$ the singular pseudo-symplectic group is introduced and for which problems (i)–(iii) are studied [56].

(4) The affine classification of quadrics over finite fields is obtained [52, 53].

The foregoing results together with our results obtained in the mid-1960s are compiled in a monograph [64].

5. Application to Association Schemes and Designs

In 1952 Bose and Shimamato [4] separated the concept of association schemes from the concept of PBIB designs. Around 1960 Hsu emphasized the importance of association schemes in various occasions in China. He also made important contributions to association schemes [2]; for instance, in 1964 he defined the association scheme of the Grassmann manifold of the m-dimensional subspaces in $\mathbb{F}_q^{(n)}$ ($1 \leqslant m < n$) by defining two m-dimensional subspaces P and Q to be the ith associates if $\dim(P+Q) = m+i$ ($i=1, 2, \cdots, \min\{m, n-m\}$) and computed its parameters v, n_i, p_{jk}^i ($i, j, k = 1, 2, \cdots, \min\{m, n-m\}$). This is one of the earliest nontrivial association schemes besides the known strongly regular graphs in the early sixties. Following Hsu, in 1965 the author [49] used a group theoretical method to compute the parameters of this association scheme which seems to be simpler, and in 1965–1966 Dai, Feng, Yang, and the authors [67, 74, 75] defined association schemes of the maximal totally isotropic (or singular) subspaces in the symplectic, unitary, and orthogonal geometries over $\mathbb{F}_q$ (which was later called the dual polar spaces of type C_v, $^2A_{2v}$, $^2A_{2v-1}$, B_v, D_v, and $^2D_{v+1}$ [5]) in the same way as the association scheme of the Grassmann manifold defined by Hsu and computed their parameters v, n_i, p_{jk}^i. The parameters v are given by the Anzahl theorems, and the parameters n_i and p_{jk}^i are computed by the group-theoretical method. All these results were complied in the monograph [67].

In 1954 Clatworthy [13] showed that a geometric configuration in $PG(3, \mathbb{F}_q)$ may be interpreted as a PBIB design. In our terminology, he took the set of one-dimensional subspaces of the four-dimensional symplectic space over $\mathbb{F}_q$ as the set of points and the set of two-dimensional totally isotropic subspaces as the set of blocks. Two points are said to be the first associates (or second associates) if they span a two-dimensional totally isotropic subspace (or nonisotropic subspace), respectively. A point is defined to lie in a block if the one-dimensional subspace of the point is contained in the two-dimensional totally isotropic subspace as the block. Then a PBIB(2) design is obtained. Clatworthy also computed the parameters of the design.

In 1962 Ray-Chaudhuri [42] used the geometry of orthogonal groups, which was called the geometry of quadrics by him, to construct PBIB designs. He constructed several PBIB(2) designs with one-dimensional or two-dimensional totally isotropic (or singular) subspaces of the geometry of orthogonal groups over finite fields as points or blocks and computed their parameters. At that time only the length of the orbit of totally isotropic (or singular) subspaces of a given dimension under the orthogonal group over finite fields was known, so he naturally restricted himself to take only totally isotropic (or singular) subspaces as points and blocks in order to compute the parameters of the designs he constructed.

In the mid-1960s after we had found the closed formulas for the lengths of all the orbits of subspaces under the symplectic, unitary, and orthogonal groups over finite fields, we [48,67,69,74,75] constructed many PBIB designs by taking the one-dimensional, two-dimensional, or v-dimensional totally isotropic (or singular) subspaces in the geometries of these classical groups as points and subspaces of any given type as blocks and computed their parameters. All these results were also compiled in the monograph [67] and were sketched in [55].

However, we would like to mention that when we took the two-dimensional totally isotropic (or singular) subspaces as points to construct association schemes and PBIB designs, we consider only the symplectic, unitary, and orthogonal spaces of low dimensions, because it gave us PBIB (2) designs. Of course, we can take any orbit of subspaces under the symplectic, unitary, or orthogonal group over finite fields as the set of points and define the associate relation according to the orbit of pairs of points under that group, then an association scheme is obtained. Moreover, if we take any orbit of subspaces as the set of blocks and define a point to lie in a block in a certain way, then a PBIB design is obtained. In a short note [50] published in 1965 some association schemes and PBIB designs were constructed by taking the one-dimensional nonisotropic (or nonsingular) subspaces in the unitary or orthogonal geometry over some small fields as points and their parameters were computed. In the 1980s the ideas of taking the one-dimensional nonisotropic (or nonsingular) subspaces and taking the two-dimensional totally isotropic (or singular) subspaces as points were carried out and generalized by several Chinese mathematicians and their works were sketched in [55] and will got be repeated here.

6. Application to Authentication Codes

The first authentication code, constructed by Gilbert et al. [23] in 1974, were based on the projective geometry over finite fields. Projective geometry, according to Klein's Erlangen Program, is the geometry of the projective general linear group. Then it is natural to propose the problem whether it is possible to construct authentication codes from the geometry of symplectic, unitary, or orthogonal groups over finite fields. The answer is of course positive and some authentication codes have been so constructed [19, 57 58, 68]. To illustrate we give a construction of Feng and Wan [19] below.

Consider the $2v$-dimensional symplectic space over $\mathbb{F}_q$, i.e., the $2v$-dimensional row vector space $\mathbb{F}_q^{(2v)}$ on which the symplectic group $\mathrm{Sp}_{2v}(\mathbb{F}_q)$ acts. Assume that $v \geqslant 2$ and let s be an integer such that $1 \leqslant s < v$. Let P_0 be a fixed subspace of type $(s, 0)$. Take the set of subspaces of type $(2s, s)$ containing P_0 to be the set $\mathscr{S}$ of source states, the set of s-dimensional subspaces whose joins with P_0 are subspaces of type $(2s, s)$ to be the set $\mathscr{E}$ of encoding rules and also the set $\mathscr{M}$ of messages. For any source state s and encoding rule e, let $f(s, e) = s \cap e^\perp$, where

$$e^\perp = \{x \in \mathbb{F}_q^{(2v)} | xK^\mathrm{t}e = 0\}. \tag{26}$$

It can be proved that $s \cap e^\perp$ is an s-dimensional subspace whose join with P_0 is of type $(2s, s)$. Thus, we may define $f(s, e) = s \cap e^\perp$ to be the message into which the source state s is encoded using the encoding rule e. Then a Cartesian authentication code is obtained and its size parameters are

$$|\mathscr{S}| = q^{2s(v-s)}, \quad |\mathscr{E}| = |\mathscr{M}| = q^{s(2v-s)}. \tag{27}$$

Now assume that the encoding rules are chosen according to a uniform probability distribution. Then the probabilities of a successful impersonation attack and a successful substitution attack are, respectively,

$$P_{\mathrm{I}} = \frac{1}{q^{s^2}}, \quad P_{\mathrm{S}} = \frac{1}{q^s}. \tag{28}$$

In virtue of the combinatorial lower bounds $P_{\mathrm{I}} \geqslant |\mathscr{S}|/|\mathscr{M}|$ and $P_{\mathrm{S}} \geqslant (|\mathscr{S}|-1)/(|\mathscr{M}|-1)$, for the authentication code constructed above P_{I} is optimal. If we require the order of magnitude of P_{S} as a function of q to be optimal, then for the code, P_{S} is nearly optimal when and only when $s = 1$.

Similar constructions can be done for the unitary and orthogonal cases.

7. Application to Projective Codes

Let C be a linear $[n, k]$-code over $\mathbb{F}_q$. C is called a *projective code* if the columns of a generator matrix of C are not proportional.

Let Q be a $k \times k$ matrix over $\mathbb{F}_q$. The set of points ${}^1(x_1, x_2, \cdots, x_k)$ in PG $(k-1, \mathbb{F}_q)$ satisfying

$$(x_1, x_2, \cdots, x_k) Q^{\mathrm{t}}(x_1, x_2, \cdots, x_k) = 0 \tag{29}$$

is called a *quadric* in PG $(k-1, \mathbb{F}_q)$ and will also be denoted by Q. The number of points of Q will be denoted by $|Q|$ and can be derived from the Anzahl theorems of the singular orthogonal geometry over $\mathbb{F}_q$. In particular, if Q is nondegenerate, i.e., (29) cannot be carried under projective transformations into a quadratic homogeneous equation with less than k indeterminates, then $|Q|$ was obtained by Primrose [40] and is also contained in Segre's formula (5).

Now let Q be a quadric in PG $(k-1, \mathbb{F}_q)$ and assume that $|Q| = n$. For each point Q choose a system of coordinates and regard it as a k-dimensional column vector. Arrange these n column vectors in any order into a $k \times n$ matrix, denoted by G_Q. It can be proved that G_Q is of rank k. Hence, G_Q can be regarded as a generator matrix of a projective $[n, k]$-code, which is denoted by C_Q and called the projective code from the quadric Q in PG $(k-1, \mathbb{F}_q)$.

In 1975 Wolfmann [73] proved that if Q is nondegenerate, then C_Q has only two or three distinct nonzero weights when k is even or odd, respectively, by computing the values of the nonzero weights of C_Q, and he computed also their multiplicities.

Let C be a linear $[n, k]$-code over $\mathbb{F}_q$. For a linear subcode D of C, let $\chi(D)$ be the *support* of D, namely,

$$\chi(D) = \{i | x_i \neq 0 \text{ for some } (x_1, x_2, \cdots, x_n) \in D\}. \tag{30}$$

The rth *minimum support weight* [24] or *generalized Hamming weight* [70] of C, denoted by $d_r(C)$, is then defined as

$$d_r(C) = \min\{|\chi(D)| | D \text{ is an } r\text{-dimensional subcode of } C\}. \tag{31}$$

Obviously, $d_1(C)$ is just the minimum Hamming weight of the code C. The *weight hierarchy* of C is then defined to be the set of generalized Hamming weights

$$\{d_1(C), d_2(C), \cdots, d_k(C)\}. \tag{32}$$

Using our results on the geometry of orthogonal groups over finite field, the author [65] computed the weight hierarchies of the q-ary projective codes from nondegenerate quadrics in projective spaces.

The corresponding problems for the projective codes from nonsingular Hermitian varieties in projective spaces over $\mathbb{F}_{q^2}$ were studied by Chakravarti [9], and Hirschfeld et al. [25]. More precisely, the former proved that the code has only two distinct nonzero weights and computed the values and multiplicities of them, and the latter three computed the weight hierarchies of the code.

A linear $[n, k]$-code C is said to satisfy the *chain condition* [71], if there exist r-dimensional subcodes D_r of C for $1 \leqslant r \leqslant k$ such that

$$|\chi(D_r)| = d_r(C), \quad r = 1, 2, \cdots, k, \qquad (33)$$

and

$$D_1 \subseteq D_2 \subseteq \cdots \subseteq D_k. \qquad (34)$$

Using the geometry of orthogonal groups over finite fields, the author [65] proved that the projective codes C_Q from nondegenerate quadrics Q in $\mathrm{PG}(k-1, \mathbb{F}_q)$ satisfy the chain condition.

8. Application to Lattices Generated by Orbits of Subspaces

Let G_n denote any one of the classical groups $\mathrm{GL}_n(\mathbb{F}_q)$, $\mathrm{Sp}_{2v}(\mathbb{F}_q)(n = 2v)$, $U_n(\mathbb{F}_q)(q = q_0^2)$, $O_{2v+\delta}(\mathbb{F}_q)(n = 2v+\delta, \delta = 0, 1, \text{or } 2)$, and $\mathrm{Ps}_{2v+\delta}(\mathbb{F}_q)$ (q even, $n = 2v+\delta, \delta = 1$ or 2), $\mathscr{M}$ be any orbit of subspaces of $\mathbb{F}_q^{(n)}$ under G_n, and $\mathscr{L}$ be the set of intersections of elements of $\mathscr{M}$. Partially order $\mathscr{L}$ by the reverse inclusion, then $\mathscr{L}$ is a finite geometric lattice with $\mathbb{F}_q^{(n)}$ as its minimal element, $\mathscr{M}$ as its set of atoms, and $r(X) = \dim(\mathbb{F}_q^{(n)}/X)$ for any $X \in \mathscr{L}$ as its rank function. $\mathscr{L}$ is called the lattice generated by $\mathscr{M}$ and the characteristic polynomial of $\mathscr{L}$ is defined by

$$\chi(\mathscr{L}, t) = \sum_{X \in \mathscr{L}} \mu(\mathbb{F}_q^{(n)}, X) t^{\dim X}, \qquad (35)$$

where μ is the Möbius function of $\mathscr{L}$. For any fixed G_n, the following problems can be proposed. (i) What are the inclusion relations between the different $\mathscr{L}$'s? (ii) What are the conditions for a subspace belonging to a given $\mathscr{L}$? (iii) The determination of the characteristic polynomials of all the $\mathscr{L}$'s. When $G_n = O_{2v+\delta}(\mathbb{F}_q)$ or $\mathrm{Ps}_{2v+\delta}(\mathbb{F}_q)$, let $\mathscr{M}'$ be any set of subspaces of the same dimension and rank. For $O_{2v+\delta}(\mathbb{F}_q)$, q being odd, let S be the $(2v+\delta) \times (2v+\delta)$ nonsingular symmetric matrix defining the group $O_{2v+\delta}(\mathbb{F}_q)$, the rank of a subspace P of $\mathbb{F}_q^{(2v+\delta)}$ is defined to be rank PS^tP. For $O_{2v+\delta}(\mathbb{F}_q)$ and $\mathrm{Ps}_{2v+\delta}(\mathbb{F}_q)$, q being even, the rank of a subspace can be defined in a similar way. Let $\mathscr{L}'$ he the set of intersections of elements of $\mathscr{M}'$. Partially order $\mathscr{L}'$ by the reverse inclusion. then $\mathscr{L}'$ is also a finite-geometric lattice and the same problems (i)–(iii) can be proposed for the $\mathscr{L}'$s.

In 1985 Orlik and Solomon [37] gave the solutions to problems (ii) and (iii) for the case when $G_n = U_n(\mathbb{F}_q)(q = q_0^2)$ and $\mathscr{M}$ is the orbit of $(n-1)$-dimensional nonisotropic subspaces and for the case when $G_n = O_{2v+\delta}(\mathbb{F}_q)$, ($q$ odd), and $\mathscr{M}'$ is the set of $(n-1)$-dimensional nonisotropic subspaces. In 1991 and 1993, respectively, Chen and Wan [10,11] gave the solutions to problems (ii) and (iii) for the case when $G_n = \mathrm{Sp}_{2v}(\mathbb{F}_q)$ and $\mathscr{M}$ is the orbit of $(2v-2)$-dimensional nonisotronic subspaces and for the case when $G_n = O_{2v+\delta}(\mathbb{F}_q)$, ($q$ even), and $\mathscr{M}'$ is the set of $(n-1)$-dimensional nonsingular subspaces. Then, for any G_n, lattices generated by the $\mathscr{M}$'s are studied by Huo et al. [27–29, 32] and lattices generated by the $\mathscr{M}'$'s are studied by Huo and Wan [30, 31, 33]. Complete solutions of problems (i)–(iii) are obtained except some boundary cases when $G_n = U_n(\mathbb{F}_q)(q = q_0^2)$, $O_{2v+\delta}(\mathbb{F}_q)(n = 2v+\delta)$ and $\mathrm{Ps}_{2v+\delta}(\mathbb{F}_q)$ (q even), which will not be described here.

We again illustrate our study with the symplectic case [27]. Denote by $\mathscr{M}(m,s;2v)$ the orbit of subspaces of type (m,s) in the $2v$-dimensional symplectic geometry over $\mathbb{F}_q$ and by $\mathscr{L}(m,s;2v)$ the lattice generated by $\mathscr{M}(m,s;2v)$. Our main results are as follows:

(i) Assume that $2s \leqslant m \leqslant v+s$ and $2s_1 \leqslant m_1 \leqslant v+s_1$, then $\mathscr{L}(m,s;2v) \supset \mathscr{L}(m_1,s_1;2v)$ if and only if $m - m_1 \geqslant s - s_1 \geqslant 0$.

(ii) Assume that $2s \leqslant m \leqslant v+s$, then $\mathscr{L}(m,s;2v)$ consists of $\mathbb{F}_q^{(2v)}$ and all subspaces of type (m_1,s_1) with $m - m_1 \geqslant s - s_1 \geqslant 0$.

(iii) Assume that $2s \leqslant m \leqslant v+s$ and $m \neq 2v$, then

$$\chi(\mathscr{L}(m,s;2v),t)$$

$$= \sum_{s_1=s+1}^{v} \sum_{m_1=2s_1}^{v+s_1} N(m_1, s_1; 2v)(t-1)(t-q)\cdots(t-q^{m_1-1})$$

$$+ \sum_{s_1=0}^{s} \sum_{m_1=m-s+s_1+1}^{v+s_1} N(m_1, s_1; 2v)(t-1)(t-q)\cdots(t-q^{m_1-1}), \quad (36)$$

where $N(m_1, s_1; 2v)$ is given by (10).

9. Application to Representations of Forms by Forms

Let us introduce some notations. Let A be an $m \times m$ matrix and B be a $t \times t$ matrix, both over $\mathbb{F}_q$.

(a) When both A and B are alternate, let

$$n_{m \times t}^{(a)}(A, B) = |\{X : m \times t |\ ^t XAX = B\}|. \quad (37)$$

(b) If q is a perfect square, say $q = q_0^2$, let $a \mapsto \bar{a} = a^{q_0}$ be the involution of $\mathbb{F}_q$, then when both A and B are hermitian, let

$$n_{m \times t}^{(h)}(A, B) = |\{X : m \times t |\ ^t \overline{X}AX = B\}|. \quad (38)$$

(c) When q is odd, and both A and B are symmetric, let

$$n_{m \times t}^{(q)}(A, B) = |\{X : m \times t |\ ^t XAX = B\}|, \quad (39)$$

When q is even, let

$$n_{m \times t}^{(q)}(A, B) = |\{X : m \times t |\ ^t XAX \equiv B\}|, \quad (40)$$

where $^t XAX \equiv B$ means that $^t XAX - B$ is alternate.

(d) When q is even, and both A and B are symmetric, let

$$n_{m \times t}^{(s)}(A, B) = |\{X : m \times t |\ ^t XAX = B\}|. \quad (41)$$

(e) Finally, for any $m \times n$ matrix A and $t \times u$ matrix B, both over $\mathbb{F}_q$, let

$$n_{m \times t, n \times u}^{(b)}(A, B) = |\{(X, Y), X : m \times t, Y : n \times u |\ ^t XAY = B\}|. \quad (42)$$

Clearly, $n_{m \times t}^{(a)}(A, B)$ is the number of representations of the alternate form with coefficient matrix B by the alternate form with coefficient matrix A, etc.

Assuming that q is odd, in 1954 Carlitz [6, 7] obtained closed formula for $n_{m \times t}^{(a)}(A, B)$ and $n_{m \times t}^{(q)}(A, B)$, and in 1955 Carlitz and Hodges [8] obtained closed formula for $n_{m \times t}^{(h)}(A, B)$. Now, using the geometry of classical groups over finite fields, or more precisely, our Anzahl theorems [64], the author [66] obtained closed formula for $n_{m \times t}^{(a)}(A, B)$, $n_{m \times t}^{(q)}(A, B)$, and $n_{m \times t}^{(h)}(A, B)$ without assuming that q is odd. Moreover, closed formulas for $n_{m \times t}^{(s)}(A, B)$ and $n_{m \times t, n \times u}^{(b)}(A, B)$ are also obtained [66].

As an illustration we sketch the deduction of $n_{m \times t}^{(a)}(A, B)$ in steps as follows:

(i) Let
$$A = \begin{pmatrix} K_{2v} & \\ & 0^{(m-2v)} \end{pmatrix}, \tag{43}$$
where K_{2v} is the $2v \times 2v$ nonsingular alternate matrix (6). Then
$$n_{m \times t}^{(a)}(A, B) = q^{t(m-2v)} n_{2v \times t}^{(a)}(K_{2v}, B). \tag{44}$$

(ii)
$$n_{2v \times t}^{(a)}(K_{2v}, B) = \begin{cases} 0 & \text{if } v_1 > v, \\ n_{2v \times 2v_1}^{(a)}(K_{2v}, K_{2v_1}) n_{2(v-v_1) \times (t-2v_1)}^{(a)}(K_{2(v-v_1)}, 0^{(t-2v_1)}) & \text{if } v_1 \leq v. \end{cases} \tag{45}$$

(iii) Let $v_1 \leq v$, then
$$n_{2v \times 2v_1}^{(a)}(K_{2v}, K_{2v_1}) = q^{v_1(2v-v_1)} \prod_{i=v-v_1+1}^{v} (q^{2i} - 1). \tag{46}$$

(iv)
$$n_{2v \times t}^{(a)}(K_{2v}, 0^{(t)}) = \sum_{k=0}^{\min\{t,v\}} q^{\binom{k}{2}} \frac{\prod_{i=t-k+1}^{t}(q^i - 1) \prod_{i=v-k+1}^{v}(q^{2i} - 1)}{\prod_{i=1}^{k}(q^i - 1)} \tag{47}$$

Formulas in Steps (i) and (ii) are easy, and formulas in Steps (iii) and (iv) can be deduced from the Anzahl theorems in symplectic geometry. Finally, combining the formula for $n_{2v \times t}^{(a)}(K_{2v}, 0^{(t)})$ obtained by Carlitz [6] with (iv) we have the following identity:

$$\sum_{k=0}^{\min\{l,v\}} q^{\binom{k}{2}} \frac{\prod_{i=t-k+1}^{t}(q^i - 1) \prod_{i=v-k+1}^{v}(q^{2i} - 1)}{\prod_{i=1}^{k}(q^i - 1)}$$

$$= q^{-\binom{t}{2}+2vt} {}_2\Phi_0[q^{-\frac{1}{2}t}, q^{-\frac{1}{2}(t-1)}; q^{v+1}]', \tag{48}$$

where

$${}_2\Phi_0[a, b; z] = \sum_{k=0}^{\infty} \frac{(a)_k (b)_k}{(q)_k} z^k,$$

$(a)_k = (1-a)(1-qa)\cdots(1-q^{k-1}a)$ if $k \geqslant 1$,

$(a)_0 = 1$,

and the prime indicates that q is to be replaced by q^{-2} in the function ${}_2\Phi_0$.

10. Concluding Remarks

Besides the applications mentioned above it should be added that the geometry of classical groups was also used in the study of correlation properties of binary m-sequences [21,22,34,35].

Classical groups are important algebraic structures and orbits of subspaces are basic geometric objects. Therefore, it is not strange that they have so many applications to the various problems mentioned above and it is natural to hope that further applications will be found in the future.

References

[1] C. Arf, Untersuchungen über quadratischen formen in körpern der charakteristik 2, Teil I, J. Reine Angew. Math. 183 (1941) 148–167.

[2] Bancheng, Partially balanced incomplete block designs, Progress in Mathematics 7 (1964) 240–281 (in Chinese); English translation: Pao-Lu Hsu, Collected

Papers (Springer, Berlin, 1983) 473–524.

[3] R.C. Bose and 1.M. Chakravati, Hermitian varieties in a finite projective space PG (N, q^2), Canad. J. Math. 18 (1966) 1161–1182.

[4] R.C. Bose and T. Shimamato, Classification and analysis of partially balanced incomplete block designs with two associate classes, J. Amer. Statist. Assoc. 47 (1952) 151–184.

[5] P.J. Cameron, Dual polar spaces, Geom. Dedicata 12 (1982) 75–85.

[6] L. Carlitz, Representations by quadratic forms in a finite field, Duke Math. J. 21 (1954) 123–137.

[7] L. Carlitz, Representations by skew forms in a finite field, Arch. Math. 5 (1954) 19–31.

[8] L. Carlitz and J.H. Hodges, Representations by hermitian forms in a finite field, Duke Math. J. 22 (1955) 393–405.

[9] I. M. Chakravarti, Families of codes with few distinct weights from singular and non-singular Hermition varieties and quadrics in projective geometries and Hadamard difference sets and designs associated with two-weight codes, IMA Volumes in Math. and Its Applications, vol. 20 (Springer, Berlin, 1990) 35–50.

[10] D. Chen and Z. Wan, The characteristic polynomial of a geometric lattice in symplectic geometry over finite fields, Chinese Sci. Bull. 36 (1991)974–978.

[11] D. Chen and Z. Wan, An arrangernent in orthogonal geometry over finite fields of char.=2, Acta Math. Sinica (N. S.) 9 (1993) 39–47.

[12] B. Chor, O. Goldreich, J. Hastad, J. Friedmann, S. Rudish and R. Smolesky, The bit extraction problem of t-resilient functions, in: Proc. 26th Symp. on Foundations of Computer Science (1985) 396–407.

[13] W.H. Clatworthy, A geometrical configuration which is a partially balanced incomplete block design, Proc. Amer Math. Soc. 5 (1954) 47–55.

[14] Z. Dai, On transitivity of subspaces in orthogonal geometry over fields of characteristic 2, Acta Math. Sinica 16 (1966) 545–560 (in Chinese); English translation: Chinese Math. 8 (1966) 569–584.

[15] Z. Dai and X. Feng, Notes on finite geometries and the construction of PBIB designs IV, Some 'Anzahl' theorems in orthogonal geometry over finite fields of characteristic not 2, Acta Sci. Sinica 13 (1964) 2001-2004. Also, Studies in finite geometries and the construction of incomplete block designs IV, Some

'Anzahl' theorems in orthogonal geometry over finite fields of characteristic $\neq 2$, Acta Math. Sinica 15 (1965) 545–558 (in Chinese); English translation: Chinese Math. 7 (1965) 265–280.

[16] L. E. Dickson, Linear Groups (Teubner, Leipzig, 1900).

[17] J. Dieudonné, Sur les groupes classiques (Hermann, Paris, 1948).

[18] J. Dieudonné, On the structure of unitary groups, Trans. Amer. Math. Soc. 72 (1952) 367–385.

[19] R. Feng and Z. Wan, A construction of Cartesian authentication codes from geometry of classical groups, J. Combin. Inform. System Sci. 20 (1995) 197–210.

[20] X. Feng and Z. Dai, Notes on finite geometries and the construction of PBIB designs V, Some 'Anzahl' theorems in orthogonal geometry over finite fields of characteristic 2, Acta Sci. Sinica 13 (1964) 2005–2008, Also, Studies in finite geometries and the construction of incomplete block designs V, Some 'Anzahl' theorems in orthogonal geometry over finite fields of characteristic 2, Acta Math. Sinica 15 (1965) 664–682 (in Chinese); English translation: Chinese Math. 7 (1965) 392–410.

[21] R.A. Games, The geometry quadrics and correlations of sequences, IEEE Trans. Inform. Theory IT-32 (1986) 423–426.

[22] R.A. Games, The geometry of m-sequences: three-valued crosscorrelations and quadrics in finite projective geometry, SIAM J. Algebraic Discrete Math. 7 (1986) 43–52.

[23] E.N. Gilbert, F.J. MacWilliams and N.J.A. Sloane, Codes which detect deception, Bell System Technical J. 53 (1974) 405–424.

[24] T. Helleseth, T. Kløve and J. Mykkeltveit, The weight distribution of irreducible cyclic codes with block length $n_1((q^l-1)/N)$, Discrete Math. 18 (1977) 179–211.

[25] J.W.P. Hirschfeld, M.A. Tafasman and S.G. Vladut, The weight hierarchy of higher-dimensional Hermitian codes, IEEE Trans. Inform. Theory IT-40 (1994) 275–278.

[26] L.K. Hua, A generalization of Hermitian matrices, Acta Sci. Sinica 2 (1953) 1–58.

[27] Y. Huo, Y. Liu and Z. Wan, Lattices generated by transitive sets of subspaces under finite classical groups Ⅰ, Comm. Algebra 20 (1992) 1123–1144.

[28] Y. Huo, Y. Liu and Z. Wan, Lattices generated by transitive sets of subspaces under finite classical groups II, Comm. Algebra 20 (1992) 2685–2727.

[29] Y. Huo, Y. Liu and Z. Wan, Lattices generated by transitive sets of subspaces under finite classical groups III, Comm. Algebra 21 (1993) 2351–2393.

[30] Y. Huo and Z. Wan, Lattices generated by subspaces of the same dimension and rank in orthogonal geometry over finite fields of odd characteristic, Comm. Algebra 21 (1993) 4219–4252.

[31] Y. Huo and Z. Wan, Lattices generated by subspaces of the same dimension and rank in orthogonal geometry over finite fields of even characteristic, Comm. Algebra 22 (1994) 201–224.

[32] Y. Huo and Z. Wan, Lattices generated by transitive sets of subspaces under finite pseudo-symplectic groups, Communications in Algebra 23 (1995) 3753–3777.

[33] Y. Huo and Z. Wan, Lattices generated by subspaces of the same dimension and rank in finite pseudo-symplectic spaces, Communications in Algebra 23 (1995) 3779–3798.

[34] T. Høholdt and J. Justeen, Tenary sequences with perfect periodic autocorrelation, IEEE Trans. Inform. Theory IT-19 (1983) 597–600.

[35] W.A. Jackson and P.R. Wild, Relations between two perfect sequence constructions, Designs Codes Cryptography 2 (1992) 325–332.

[36] Y. Liu and Z. Wan, Pseudo–symplectic geometries over finite fields of characteristic two, in: J Hirschfeld et al., eds., Recent Advances on Finite Geometries and Designs (Oxford University Press, Oxford. 1991) 265–288.

[37] P. Orlick and L. Solomon, Arrangements in unitary and orthogonal geometry over finite fields, J. Combin. Theory Ser. A 38 (1985) 217–229.

[38] V. Pless, On Witt's theorem for nonalternating symmetric bilinear forms over a field of characteristic 2, Proc. Amer. Math. Soc. 15 (1964) 979–983.

[39] V. Pless, The number of isotropic subspaces in a finite geometry, Atti. Accad. Naz. Lincei. Rend. (8) 39 (1965) 418-421.

[40] E.J.F. Primrose, Quadrics in finite geometries, Proc. Cambridge Philos. Soc. 47 (1951)299–304.

[41] D. K. Ray-Chaudhuri, Some results on quadrics in finite projective geometry based on Galois fields, Canad. J. Math. 14 (1962) 129–138.

[42] D.K. Ray-Chaudhuri, Application of the geometry of quadrics for constructing

PBIB designs, Ann Math. Statist. 33 (1962) 1175–1186.

[43] B. Segre, On Galois geometries, Proc. Internat. Congress Math., Edinburgh (1958) (Cambridge Univ Press, Cambridge, 1960) 488–499.

[44] B. Segre, Le geometrie di Galois, Ann. Math. Pura Appl. 48 (1959) 1–97.

[45] B. Segre, Forme e geometrie hermitiane con particolare riguardo al caso finito, Ann. Mat. Pura Appl (4) 70 (1965) 1–201.

[46] G. Simmons, Authentication theory/coding theory, Advances in Cryptology, in: Proc. of Crypto 84 Lecture Notes in Computer Science, vol. 196 (Springer, Berlin, 1985) 411–431.

[47] Z. Wan, Notes on finite geometries and the construction of PBIB designs I, Some 'Anzahl' theorems in symplectic geometry over finite fields, Acta Sci. Sinica 13 (1964) 515–516; Studies in finite geometries and the construction of incomplete block designs I, Some 'Anzahl' theorems in symplectic geometry over finite fields, Acta Math. Sinica 15 (1965) 354–361 (in Chinese); English translation: Chinese Math. 7 (1965) 55–62.

[48] Z. Wan, Notes on finite geometries and the construction of PBIB designs II, Some PBIB designs with two associate classes based on the symplectic geometry over finite fields, Sci. Sinica 13 (1964) 516–517; Studies in finite geometries and the construction of incomplete block designs II Some PBIB designs with two associate classes based on symplectic geometry over finite fields, Acta Math. Sinica 15 (1965) 362–371 (in Chinese); English translation: Chinese Math. 7 (1965) 63–72.

[49] Z. Wan, Studies in finite geometries and the construction of incomplete block designs VI, Some association schemes with several associate classes based on subspaces of a vector space over finite field, Shuxue Jinzhan 8 (1965) 293–302 (in Chinese).

[50] Z. Wan, Notes on finite geometries and the construction of PBIB designs VI, Some association schemes and PBIB designs based on finite geometries, Acta Sci. Sinica 14 (1965) 1872–1876.

[51] Z. Wan, On the symplectic invariants of a subspace of a vector space, Acta Math. Sci. 11 (1991) 251–253.

[52] Z. Wan, Quadrics in $\mathrm{AG}(n, \mathbb{F}_q)$ for q odd, Chinese Sci. Bull. 36 (1991) 2014–2015.

[53] Z. Wan, Quadrics in $\mathrm{AG}(n, \mathbb{F}_q)$ for q even, Chinese Sci. Bull. 36 (1991) 2016–

2017.

[54] Z. Wan, On the unitary invariants of a subspace of a vector space over a finite field, Chinese Sci. Bull 37 (1992) 705–707.

[55] Z. Wan, Finite geometries and block designs, Sankhyā: The Indian J. Statist., Special volume 54 (1992) 531–543.

[56] Z. Wan, Singular pseudo-symplectic geometry over finite fields of characteristic 2, Northeastern Math. 8 (1992) 391–416.

[57] Z. Wan, Further constructions of Cartesian authentication codes from symplectic geometry, Northeastern Math. J. 8 (1992) 4–20.

[58] Z. Wan, Construction of Cartesian authentication codes from unitary geometry, Designs, Codes, Cryptography 2 (1992) 333–356.

[59] Z. Wan, Introduction to Abstract and Linear Algebra (Studentlitteratur, Lund, 1992).

[60] Z. Wan, On the orthogonal invariants of a subspace of a vector space over a finite field of odd characteristic, Linear Algebra Appl. 184 (1993) 123–133.

[61] Z. Wan, On the orthogonal invariants of a subspace of a vector space over a finite field of even characteristic, Linear Algebra Appl. 184 (1993) 135–143.

[62] Z. Wan, Some Anzahl theorems in finite singular symplectic, unitary and orthogonal geometries, Discrete Math. 123 (1993) 131–150.

[63] Z. Wan, Further studies on singular symplectic, unitary, and orthogonal geometries over finite fields, Southeast Asian Bull. Math. 17 (1993) 177–196.

[64] Z. Wan, Geometry of Classical Groups over Finite Fields (Studentlitteratur, Lund, 1993).

[65] Z. Wan, The weight hierarchies of the projective codes from nondegenerate quadrics, Designs, Codes Cryptography 4 (1994) 283–300.

[66] Z. Wan, Representations of forms by forms in a finite field, Finite Fields and Their Applications 1(1995) 297–325.

[67] Z. Wan, Z. Dai, X. Feng and B. Yang, Studies on Finite Geometry and the Construction of Incomplete Block Designs (Science Press, Beijing, 1966) (in Chinese).

[68] Z. Wan, B. Sweets and P. Vanroose, On the construction of authentication codes over symplectic spaces, IEEE Trans. Inform. Theory IT-40 (1994) 920–929.

[69] Z. Wan and B. Yang, Notes on finite geometries and the construction of PBIB designs III, Some 'Anzahl' theorems in unitary geometry over finite fields and

their applications, Acta Sci. Sinica 13 (1964) 1006–1007. Also, Studies in finite geometries and the construction of incomplete block designs III, Some 'Anzahl' theorems in unitary geometry over finite fields and their applications, Acta Math Sinica 15 (1965) 533–544 (in Chinese); English translation: Chinese Math. 7 (1965) 252–264.

[70] V.K. Wei, Generalized Hamming weights for linear codes, IEEE Trans. Inform. Theory IT-37 (1991) 1412–1418.

[71] V.K. Wei and K. Yang, On the generalized Hamming weights of squaring construction and product codes, IEEE Trans. Inform. Theory 39 (1993) 1707–1713.

[72] E. Witt, Theorie der quadratischen Formen in beliebigen Körpern, J. Reine Angew. Math. 176 (1937) 31–44.

[73] J. Wolfman, Codes projectifs à deux ou trois poids associés aux hyperquadriques d'une géométrie finie Discrete Math. 13 (1975) 185–211.

[74] B. Yang, Studies in finite geometries and the construction of incomplete block designs VII, Association schemes with several associate classes by taking the maximal totally isotropic subspaces in the symplectic geometry over finite fields as treatments, Acta Math. Sinica 15 (1965) 812–825 (in Chinese); English translation: Chinese Math. 7 (1965) 547–560.

[75] B. Yang, Studies in finite geometries and the construction of incomplete block designs VIII, Association schemes with several associate classes by taking the maximal totally isotropic subspaces in the unitary geometry over finite fields as treatments, Acta Math. Sinica 15 (1965) 826–841 (in Chinese); English translation: Chinese Math. 7 (1965) 561–576.

Nonlinear Feedforward Sequences of m-sequences I*

Zongduo Dai, Xuning Feng,
Mulan Liu and Zhexian Wan

Academia Sinica, Beijing, China

Abstract The minimal polynomial of a nonlinear feedforward sequence of an m sequence is studied, and algorithms for synthesising the nonlinear feedforward sequence are obtained.

1. Introduction

Let
$$\alpha = (a_0, a_1, \cdots, a_t, \cdots), \quad a_t \in \mathbb{F}_2$$
be a given n-stage m-sequence over the binary field $\mathbb{F}_2$. Denote
$$s_i = (a_i, \cdots, a_{i+n-1}), \quad i \geqslant 0.$$
Let $\phi : \mathbb{F}_{2^n} \to \mathbb{F}_2$ be a Boolean function of n variables. Define a sequence
$$\phi(\alpha) = (\phi(s_0), \phi(s_1), \cdots, \phi(s_t), \cdots)$$
called a feedforward sequence of α by ϕ, and ϕ is then called a feedforward transformation. Since $s_t \neq 0$ for all $t \geqslant 0$, the value $\phi(\mathbf{0})$ does not effect the forward sequence $\phi(\alpha)$, we shall always assume $\phi(\mathbf{0}) = 0$.

Correspondence to: Zhexian Wan, Institute of Systems Science, Academia Sinica, Beijing 100080, China

* Presented at the Beijing International Workshop on Information Theory, 4-7 July 1988.

It is well known that a Boolean Function of n variables can be represented by a polynomial $f(x_0, x_1, \cdots, x_{n-1}) \in \mathbb{F}_2[x_0, x_1, \cdots, x_{n-1}]$, i.e.

$$\phi(a_0, a_1, \cdots, a_{n-1}) = f(a_0, a_1, \cdots, a_{n-1}) \text{ for all} (a_0, a_1, \cdots, a_{n-1}) \in \mathbb{F}_{2^n}.$$

Moreover, if we assume

$$\deg_{x_i} f(x_0, x_1, \cdots, x_{n-1}) \leqslant 1 \text{for all } i = 0, 1, \cdots, n-1, \qquad (1)$$

then the polynomial $f(x_0, x_1, \cdots, x_{n-1})$ corresponding to the given ϕ is uniquely determined. Thus, Boolean functions of n variables can be understood as polynomials satisfying (1).

Let

$$\Phi = \{\phi(x_0, x_1, \cdots, x_{n-1}) | \phi(x_0, x_1, \cdots, x_{n-1}) \in \mathbb{F}_2[x_0, x_1, \cdots, x_{n-1}],$$
$$\deg_{x_i} \phi(x_0, x_1, \cdots, x_{n-1}) \leqslant 1, \quad \phi(\mathbf{0}) = 0\}.$$

Then Φ is just the set of all feedforward transformations. Obviously

$$|\Phi| = 2^{2^n - 1}.$$

For $\phi \in \Phi$, call

$$\deg \phi(x_0, x_1, \cdots, x_{n-1})$$

the degree of the feedforward transformation. Put

$$\mathscr{S}_r = \{\phi(\alpha) | \phi \in \Phi, \deg \phi \leqslant r\}.$$

It is obvious that $\mathscr{S}_r$ is a vector space over $\mathbb{F}_2$, and for every $\phi \in \Phi$, we have $\phi(\alpha) \in \mathscr{S}_n$.

2. Preparations

Denote by $\mathscr{S}$ the set of all sequences over $\mathbb{F}_2$. We can give $\mathscr{S}$ a $\mathbb{F}_2[x]$-module structure, if addition and module operation in $\mathscr{S}$ are defined as follows:

Let

$$\alpha = (a_0, a_1, \cdots, a_t, \cdots), \quad \beta = (b_0, b_1, \cdots, b_t, \cdots) \in \mathscr{S},$$

where $a_t, b_t \in \mathbb{F}_2$ for all $t \geqslant 0$ and let

$$f(x) = \sum_{i=0}^{m} c_i x^i \in \mathbb{F}_2[x],$$

then define

$$\alpha + \beta = (a_0 + b_0, \cdots, a_t + b_t, \cdots)$$

$$f(x) * \alpha = \left(\sum_{i=0}^{m} c_i a_i, \sum_{i=0}^{m} c_i a_{1+i}, \cdots, \sum_{i=0}^{m} c_i a_{t+i}, \cdots \right).$$

In particular, x is the left translation operator.

Lemma 2.1 $\mathscr{S}$ *is an* $\mathbb{F}_2[x]$*-module.*

For any sequence α, regarded as an element of the module $\mathscr{S}$, its annihilating ideal will be denoted by Ann α, i.e.

$$\text{Ann}\,\alpha = \{f(x) | f(x) \in \mathbb{F}_2[x], f(x) * \alpha = 0\}$$

and the generator of Ann α is called the minimal polynomial of α.

Put

$$\Omega = \{\alpha | \alpha \in \mathscr{S}, \alpha \text{ being periodic}\}.$$

Obviously, Ω is a $\mathbb{F}_2[x]$-submodule of $\mathscr{S}$.

Lemma 2.2 *Let $f(x)$ be the minimum polynomial of α, then $\alpha \in \Omega$ if and only if $f(0) = 1$.*

Proof (Only if) Denote the period of α by T. Then $x^T + 1 \in \text{Ann}\,\alpha$. Since $f(x) | x^T + 1$, there exists $g(x) \in \mathbb{F}_2[x]$ such that $f(x)g(x) = x^T + 1$. Thus, $f(0)g(0) = 1$ and hence $f(0) = 1$.

(If) Since $f(0) = 1$, $(x'f(x)) = 1$ and thus there exists a positive integer T such that

$$x^T \equiv 1 (\bmod f(x)).$$

Therefore, $x^T + 1 \in \text{Ann}\,\alpha$, $x^T * \alpha = \alpha$, and hence $\alpha = \Omega$. $\square$

For any $f(x) \in \mathbb{F}_2[x]$, put

$$\Omega(f) = \{\alpha | \alpha \in \mathscr{S}, f(x) * \alpha = 0\}.$$

Clearly, we have the following lemma.

Lemma 2.3 $\Omega(f)$ is an $\mathbb{F}_2[x]$-submodule of $\mathscr{S}$.

By Lemma 2.1, we have

$$\Omega = \bigcup_{\substack{f \in \mathbb{F}_2[x] \\ f(0)=1}} \Omega(f).$$

In the following we shall describe the generating function of an element $\alpha \in \Omega$ by an element of the field of formal Laurent series $\mathbb{F}_2((x^{-1}))$ over $\mathbb{F}_2$ and give the algebraic description of the module operation and module structure of the $\mathbb{F}_2[x]$-module $\Omega(f)$.

By $\mathbb{F}_2[x]((x^{-1}))$ we mean the field of all formal Laurent series in x^{-1} over $\mathbb{F}_2$

$$\left\{ \sum_{i=m}^{\infty} a_i x^{-i} \,\Big|\, m \in \mathbb{Z}, a_i \in \mathbb{F}_2 \right\}$$

with addition and multiplication defined as follows: Let

$$A(x) = \sum_{i=m}^{\infty} a_i x^{-i}, \quad B(x) = \sum_{i=n}^{\infty} b_i x^{-i},$$

then define

$$A(x) + B(x) = \sum_{i=\min(m,n)}^{\infty} (a_i + b_i) x^{-i},$$

where we put $a_i = 0$ in case $\min(m,n) \leqslant i < m$ and $b_i = 0$ in case $\min(m,n) \leqslant i < n$, and

$$A(x) \cdot B(x) = \sum_{i=m+n}^{\infty} c_i x^{-i},$$

where

$$c_k = \sum_{i+j=k} a_i b_j.$$

It can be readily verified that $\mathbb{F}_2((x^{-1}))$ is a field with respect to addition and multiplication defined above. Moreover, we have $\mathbb{F}_2[x] \subseteq \mathbb{F}_2((x^{-1}))$, hence we also have

$$\mathbb{F}_2(x) \subseteq \mathbb{F}_2((x^{-1})).$$

Let $A(x) = \sum_{i=m}^{\infty} a_i x^{-i}$, where $a_m = 1$. Define the order of $A(x)$, denoted $o(A(x))$, to be $-m$, i.e.
$$o(A(x)) = -m.$$
And define the order of 0 to be $-\infty$, i.e. $o(0) = -\infty$.

Let $A(x) = \sum_{i=m}^{\infty} a_i x^{-i}$. Define
$$\lfloor A(x) \rfloor = \sum_{i=m}^{0} a_i x^{-i} \text{ and } \{A(x)\} = \sum_{i=1}^{\infty} a_i x^{-i},$$
and call them the integral part and fractional part of $A(x)$, respectively. Obviously, $\lfloor A(x) \rfloor \in \mathbb{F}_2[x]$. If $\lfloor A(x) \rfloor \neq 0$, then
$$\deg \lfloor A(x) \rfloor = o(A(x))$$
and
$$o(A(x)) \geq -1$$

With any $\alpha = (a_0, a_1, \cdots, a_t, \cdots) \in \mathscr{S}$, we associate a formal Laurent series
$$A(\alpha) = \sum_{i=0}^{\infty} a_i x^{-i-1} \in x^{-1} \mathbb{F}_2[[x^{-1}]] \subset \mathbb{F}_2((x^{-1}))$$
and we call $A(\alpha)$ the generating function of α. Obviously,
$$\{A(\alpha)\} = A(\alpha), \quad o(A(\alpha)) \leq -1.$$

Clearly, the map $\alpha \to A(\alpha)$ from $\mathscr{S}$ into $x^{-1}\mathbb{F}_2[[x^{-1}]]$ is linear and bijective. In the following, we usually do not distinguish α and $A(\alpha)$. Thus, A, Ω and $\Omega(f)$ can be regarded as subsets of $\mathbb{F}_2((x^{-1}))$, and then the module operation and module structure of them can be described more algebraically as follows.

Lemma 2.4 (module operation in $\mathscr{S}$) Let $\alpha \in \mathscr{S}$ and $f(x) \in \mathbb{F}_2[x]$, then
$$A(f(x) * \alpha) = \{f(x) A(\alpha)\}.$$

Proof Let $f(x), g(x) \in \mathbb{F}_2[x]$. Clearly,
$$A((f(x) + g(x)) * \alpha) = A(f(x) * \alpha) + A(g(x) * \alpha).$$

$$\{(f(x)+g(x))A(\alpha)\} = \{f(x)A(\alpha)\} + \{g(x)A(\alpha)\}.$$

Therefore, it is sufficient to prove $A(x^n * \alpha) = \{x^n A(\alpha)\}$. Suppose $\alpha = (a_0, a_1, \cdots, a_t, \cdots)$, then

$$x^n * \alpha = (a_n, a_{n+1}, \cdots, a_{n+t}, \cdots),$$

$$A(x^n * \alpha) = \sum_{i=0}^{\infty} a_{n+i} x^{-i-1},$$

$$\{x^n A(\alpha)\} = \left\{\sum_{i=0}^{\infty} a_i x^{n-i-1}\right\} = \sum_{i=0}^{\infty} a_{n+i} x^{-i-1}.$$

Thus Lemma 2.4 is proved. □

Corollary 2.5 *If we define*

$$f(x) * A(x) = \{f(x)A(x)\} \quad \forall f(x) \in \mathbb{F}_2[x], \; A(x) \in x^{-1}\mathbb{F}_2[[x^{-1}]],$$

then $x^{-1}\mathbb{F}_2[[x^{-1}]]$ is an $\mathbb{F}_2[x]$-module and $\mathscr{S}$ and $x^{-1}\mathbb{F}_2[[x^{-1}]]$ are isomorphic $\mathbb{F}_2[x]$-modules.

Lemma 2.6 (algebraic description of $\Omega(f)$)   *Suppose $f(x) \in \mathbb{F}_2[x]$, deg $f(x) = n$ and $f(0) = 1$. Then*

(i) *For any $\alpha \in \Omega(f)$, we have $f(x)A(\alpha) \in \mathbb{F}_2[x]$ and deg $(f(x)A(\alpha)) <$ deg $f(x)$. Moreover, $f(x)A(\alpha)$ can be computed in the following way: Suppose $\alpha = (a_0, a_1, \cdots, a_t, \cdots)$; if we set*

$$s_1(x) = \sum_{i=0}^{n-1} a_i x^{-i-1},$$

then we have

$$f(x)A(\alpha) = \lfloor s_1(x)f(x) \rfloor.$$

(ii) $\Omega(f) = \{[g(x)/f(x)] | g(x) \in \mathbb{F}_2[x], \deg g(x) < \deg f(x)\}.$

Proof   (i) We have

$$\begin{aligned}\{f(x)A(\alpha)\} &= A(f(x) * \alpha) \quad \text{(Lemma 2.4)} \\ &= A(0) \quad (\alpha \in \Omega(f)) \\ &= 0.\end{aligned}$$

45

Thus,
$$f(x)A(\alpha) = \lfloor f(x)A(\alpha) \rfloor \in \mathbb{F}_2[x]$$
and
$$\deg(f(x)A(\alpha)) = \mathrm{o}(f(x)A(\alpha)) = (\mathrm{o}(f) + \mathrm{o}(A(\alpha))) < \mathrm{o}(f) = \deg f.$$

Denote $s_2(x) = \sum_{i=n}^{\infty} a_i x^{-i-1}$, then
$$f(x)A(\alpha) = \lfloor f(x)A(\alpha) \rfloor = \lfloor f(x)(s_1(x) + s_2(x)) \rfloor = \lfloor f(x)s_1(x) \rfloor + \lfloor f(x)s_2(x) \rfloor.$$

But $\mathrm{o}(f(x)s_2(x)) < 0$, and hence $\lfloor f(x)s_2(x) \rfloor = 0$. Therefore,
$$f(x)A(\alpha) = \lfloor f(x)s_1(x) \rfloor.$$

(ii) For any $\alpha \in \Omega(f)$, from (i) we deduce $f(x)A(\alpha) = g(x) \in \mathbb{F}_2[x]$ and $\deg g(x) < \deg f(x)$. Thus, $\alpha = g(x)/f(x)$.

Conversely, suppose $g(x) \in \mathbb{F}_2[x]$ and $\deg g(x) < \deg f(x)$. Then
$$\mathrm{o}\left(\frac{g(x)}{f(x)}\right) = \mathrm{o}(g(x)) - \mathrm{o}(f(x)) = \deg g(x) - \deg f(x) < 0.$$

Thus, we may assume $[g(x)/f(x)] = \sum_{i=0}^{\infty} a_i x^{-i-1}$. Put
$$\alpha = (a_0, a_1, \cdots, a_t, \cdots),$$
then $[g(x)/f(x)] = A(\alpha)$. But by Lemma 2.4, we also have
$$A(f(x) * \alpha) = \{f(x)A(\alpha)\} = \{g(x)\} = 0.$$

Thus $f(x) * \alpha = 0$, i.e. $\alpha \in \Omega(f)$. $\square$

Lemma 2.7 (module operation in $\Omega(f)$) *Suppose $f(x) \in \mathbb{F}_2[x]$ and $f(0) = 1$. Let $\alpha = (g/f) \in \Omega(f)$, $h(x) \in \mathbb{F}_2[x]$. Then*
$$A(h * \alpha) = \frac{(gh)_f}{f},$$
where $(gh)_f$ denotes the remainder of gh after dividing by f, i.e. $(gh)_f \in \mathbb{F}_2[x]$, $(gh)_f \equiv gh \pmod{f}$ and $\deg(gh)_f < \deg f$.

Proof By the definition of $(gf)_f$, there exists $g(x) \in \mathbb{F}_2[x]$ such that

$$gh = gf + (gh)_f.$$

Thus

$$A(h*\alpha) = \{hA(\alpha)\} = \left\{\frac{hg}{f}\right\} = \left\{\frac{gf + (gh)_f}{f}\right\}$$
$$= \left\{g + \frac{(gh)_f}{f}\right\} = \frac{(gh)_f}{f}.$$

The first equality is by Lemma 2.4, the last equality is due to $o(q) \geqslant 0$ and $o(gh)_f/f) < 1$, and the other equalities are obvious. $\square$

Lemma 2.8 (module structure of $\Omega(f)$) *Suppose $f(x) \in \mathbb{F}_2[x]$ and $f(0) = 1$. Then*

(i) $\Omega(f) = \langle g/f \rangle$, where $g(x)$ is any polynomial in $\mathbb{F}_2[x]$, relatively prime with $f(x)$ and $\langle g/f \rangle$ denotes the cyclic submodule of Ω generated by g/f.

(ii) Ann $(g/f) = f/(f,g) \forall g/f \in \Omega(f)$.

(iii) $\Omega(f) \simeq \mathbb{F}_2[x]/(f)$ and $\dim_{\mathbb{F}_2} \Omega(f) = \deg f$.

Proof (i) Since $(f,g) = 1$, there exists a polynomial $g^{-1} \in \mathbb{F}_2[x]$ such that $g \cdot g^{-1} \equiv 1 \pmod{f}$. Then for any $g_1(x) \in \mathbb{F}_2[x]$ with $\deg g_1(x) < \deg f(x)$, putting $h(x) = g^{-1} \cdot g_1$, we have

$$g_1(x) = (gh)_f.$$

By Lemma 2.7, we have

$$\frac{g_1}{f} = h * \frac{g}{f} \in \langle g/f \rangle.$$

Threfore, by Lemma 2.6(ii) we obtain $\Omega(f) \subseteq \langle g/f \rangle$. But obviously we have $\langle g/f \rangle \subseteq \Omega(f)$. Consequently, $\Omega(f) = \langle g/f \rangle$.

(ii) By Lemma 2.7

$$\text{Ann}\left(\frac{g}{f}\right) = \{h | h \in \mathbb{F}_2[x] \ni (hg)_f = 0\}$$
$$= \{h | h \in \mathbb{F}_2[x] \ni hg \equiv 0 \pmod{f}\}$$
$$= \left\{h | h \in \mathbb{F}_2[x] \ni h \equiv 0 \left(\mod \frac{f}{(f,g)}\right)\right\}$$

$$= \left(\frac{f}{(f,g)}\right).$$

(iii) By (ii), Ann $(1/f) = (f)$. Then by (i), $\Omega(f) = \langle 1/f \rangle = \mathbb{F}_2[x]/(f)$. □

3. More Definitions

In the following we shall always assume that α is an n-stage m-sequence and $f(x) \in \mathbb{F}_2[x]$ is its minimal polynomial. It is well-known that $f(x)$ is a primitive polynomial of degree n in $\mathbb{F}_2[x]$.

At first let us introduce the concept of dyadic weights of nonnegative integers. Let $a \in \mathbb{Z}$, $a > 0$, and express a in dyadic system

$$a = \sum_{i \geqslant 0} a_i 2^i, \quad a_i = 0 \text{ or } 1.$$

Put

$$w(a) = \sum_{i \geqslant 0} a_i$$

and call it the dyadic weight of a.

Then we shall introduce the concept of weights of nonzero elements in $\mathbb{F}_{2^n}$ with respect to $f(x)$. Let θ be a root of $f(x)$ in $\bar{\mathbb{F}}_2$, the algebraic closure of $\mathbb{F}_2$, then $\mathbb{F}_2(\theta) = \mathbb{F}_{2^n}$. For every $\rho \in \mathbb{F}_{2^n}^*$, there exists a unique integer a, $0 \leqslant a < 2^n - 1$ such that

$$\rho = \theta^a.$$

Put

$$w_\theta(\rho) = w(a)$$

and call it the weight of ρ with respect to θ. If θ_1 is another root of $f(x)$, we can prove

$$w_{\theta_1}(\rho) = w_\theta(\rho).$$

In fact, there exists an integer $b \geqslant 0$ such that $\theta = \theta_1^{2^b}$. If $\rho = \theta^n$, then $\rho = \theta_1^{2^b a}$. But there exists a unique integer c, $0 \leqslant c < 2^n - 1$, such that $\rho = \theta_1^c$. Thus

$$2^b a \equiv c \pmod{2^n - 1},$$

and it follows that
$$w(c) = w(a).$$

Consequently,
$$w_{\theta_1}(\rho) = w_{\theta_1}(\theta_1^{2^b a}) = w_{\theta_1}(\theta_1^c) = w(c) = w(a).$$

Therefore, $w_\theta(\rho)$ is independent of the root of $f(x)$ chosen, and will be called the weight of ρ with respect to $f(x)$ and denoted by $w_f(\rho)$. In the following discussion the primitive polynomial $f(x)$ of degree n will be fixed, hence we may abbreviate $w_f(\rho)$ as $w(\rho)$

Now we introduce the following two sets of polynomials
$$F_r(x) = \prod_{\substack{\rho \in \mathbb{F}_{2^n}^* \\ w(\rho)=r}} (x-\rho), \quad r = 0, 1, \cdots, n$$

and
$$G_r(x) = \prod_{j=1}^{r} F_j(x), \quad r = 0, 1, \cdots, n.$$

Obviously,
$$\deg F_r(x) = \binom{n}{r},$$
$$\deg G_r(x) = \sum_{j=1}^{r} \binom{n}{j},$$
$$G_n(x) = x^{2^n - 1} + 1.$$

In Section 1, we regarded the set of feedforward transformations as a subset Φ of $\mathbb{F}_2[x_0, x_1, \cdots, x_{n-1}]$. In fact, any polynomial ψ in $\mathbb{F}_2[x_0, x_1, \cdots, x_{2^n-2}]$ can also be regarded as a feedforward transformation, whose feedforward sequence can be defined in the following way: Let the given n-stage m-sequence α be
$$\alpha = (a_0, a_1, \cdots, a_t, \cdots).$$

Put $\tilde{s}_t = (a_t, a_{t+1}, \cdots, a_{t+2^n-1})$, $t = 0, 1, 2, \cdots$. Then define
$$\psi(\alpha) = (\psi(\tilde{s}_0), \psi(\tilde{s}_1), \cdots, \psi(\tilde{s}_t), \cdots).$$

If we restrict ψ to be contained in Φ, this definition coincides with the original one given in Section 1. On the other hand, extending Φ to $\mathbb{F}_2[x_0, x_1, \cdots, x_{2^n-2}]$, we do ont obtain new feedforward transformations in essence. In other words, for any $\psi(x_0, x_1, \cdots, x_{2^n-2}) \in \mathbb{F}_2[x_0, x_1, \cdots, x_{2^n-2}]$ with $\psi(\tilde{0}) = 0$. there always exists a polynomial $\phi(x_0, x_1, \cdots, x_{n-1}) \in \Phi$ such that

$$\phi(\alpha) = \Psi(\alpha).$$

In fact, ϕ can be constructed in the following way. Let $f(x)$ be the minimal polynomial of α. Define $a_{ij} (i \geqslant 0, 0 \leqslant j < n)$ by

$$x^i \equiv \sum_{j=0}^{n-1} a_{ij} x^j \pmod{f(x)}.$$

Then take ϕ to be the polynomial satisfying

$$\phi(x_0, x_1, \cdots, x_{n-1}) \in \mathbb{F}_2[x_0, x_1, \cdots, x_{n-1}],$$

$$\phi(x_0, x_1, \cdots, x_{n-1}) \equiv \psi\left(x_0, x_1, \cdots, x_{n-1}, \sum_{j=0}^{n-1} a_{nj} x_j, \cdots, \sum_{j=0}^{n-1} a_{2^n-2\,j} x_j\right)$$
$$\pmod{x_0^2 - x_0, x_1^2 - x_1, \cdots, x_{n-1}^2 - x_{n-1}}$$

$$\deg_{x_i} \phi \leqslant 1, \ 0 \leqslant i < n.$$

Clearly, ϕ is uniquely determined and $\phi(\mathbf{0}) = 0$, thus $\phi \in \Phi$. In the following we also use polynomials in $\mathbb{F}_2[x_0, x_1, \cdots, x_{2^n-2}]$ to represent feedforward transformations.

Let $\Psi(x_0, \cdots, x_{2^n-2}) \in \mathbb{F}_2[x_0, x_1, \cdots, x_{2^n-2}]$ and $\mu(x) = \sum c_i x^i \in \mathbb{F}_2[x]$. Then a polynomial in $\mathbb{F}_2[x_0, \cdots, x_{n-1}]$, denoted by $\mu(x) * \psi$, can be determined by the following conditions:

$$\mu(x) * \psi \equiv \sum_i c_i \psi \left(\sum_{j=0}^{n-1} a_{ij} x_j, \cdots, \sum_{j=0}^{n-1} a_{i+2^n-2\,j} x_j\right)$$
$$\pmod{x_0^2 - x_0, \cdots, x_{n-1}^2 - x_{n-1}} \ \deg_{x_i}(\mu(x) * \psi) < 1.$$

Obviously, if $\psi(0) = 0$, then $\mu(x) * \psi \in \Phi$ and if deg $\psi = r$, then deg $(\mu(x) * \psi) \leqslant r$.

In the following we shall use $f(\rho, x)$, where $\rho \in \bar{\mathbb{F}}_2$, to denote the minimal polynomial of ρ over $\mathbb{F}_2$ and $f(\beta, x)$ to denote the minimal polynomial of the sequence β.

4. Main Results

The main results of the present paper read as follows:

Theorem 4.1 (i) $\mathscr{S}_r = \Omega(G_r(x))$.

(ii) Let
$$B_1(r) = \{x_{i_1} x_{i_2} \cdots x_{i_j}(\alpha) | 0 \leqslant i_1 < \cdots < i_j < n, 1 \leqslant j \leqslant r\},$$
then $B_1(r)$ is a basis of $\mathscr{S}_r$ over $\mathbb{F}_2$.

Note Using Theorem 4.1(ii) we may devise a synthesis algorithm for the feedforward sequence $\phi(\alpha)$. In fact, assume deg $\phi \leqslant r$. By Theorem 4.1(ii), there exist $c_{i_1, i_2 \cdots i_j} \in \mathbb{F}_2 (0 \leqslant i_1 < i_2 < \cdots < i_j < n, 1 \leqslant j \leqslant r)$ such that

$$\phi(\alpha) = \sum_{j=1}^{r} \sum_{0 \leqslant i_1 < \cdots < i_j < n} c_{i_1 i_2 \cdots i_j} x_{i_1} x_{i_2} \cdots x_{i_j}(\alpha).$$

Solving, we get $c_{i_1 i_2 \cdots i_j}$. Then

$$\phi = \sum_{j=1}^{r} \sum_{0 \leqslant i_1 < \cdots < i_j < n} c_{i_1 i_2 \cdots i_j} x_{i_1} x_{i_2} \cdots x_{i_j}.$$

Let

$$\phi(x_0, \cdots, x_{2^n-2}) \in \mathbb{F}_2[x_0, x_1, \cdots, x_{2^n-2}] \text{ and deg } \phi = r.$$

If $F_r(x) | f(\phi(\alpha), x)$, then ϕ is called a nondegenerate feedforward transformation of de- gree r.

Theorem 4.2 Let l be a positive integer such that $l \not\equiv 0 (\mod 2^n - 1/2^\tau - 1)$ for all $\tau | n$. Then $x_0 x_1 \cdots x_{(r-1)l}$ is a nondegenerate feedforward transformation of degree r.

Theorem 4.3 Let $1 \leqslant r < n, (r, n) = 1$ and $\phi(x_0, x_1, \cdots, x_{2^n-2}) \in \mathbb{F}_2[x_0, x_1, \cdots, x_{2^n-2}]$ with $\phi(\tilde{0}) = 0$ be a nondegenerate feedforward transformation of degree r. For every $\mu(x) \in \mathbb{F}_2[x], \mu(x) \neq 0, \deg \mu(x) < n$ and $\psi(x_0, x_1, \cdots, x_{n-1}) \in \Phi, \deg \psi < r$, then

(i) $\mu(x) * \phi + \psi$ is also nondegenerate feedforward transformation of degree r.

(ii) For $\mu_1(x) \in \mathbb{F}_2[x], \mu_1(x) \neq 0, \deg \mu_1(x) < n$, and $\psi_1(x_0, x_1, \cdots, x_{n-1}) \in \Phi, \deg \psi_1 < r, \mu * \phi + \psi = \mu_1 * \phi + \psi$, implies $\mu = \mu_1$ and $\psi = \psi_1$.

Note According to Theorem 4.3, from a given nondegenerate feedforward transformation of degree r,

$$(2^{n-1}) 2^{\Sigma_i^{r-1} \binom{n}{i}}$$

nondegenerate feedforward transformations of degree r can be derived.

Theorem 4.4 (i) Let

$$B_2(r) = \left\{ x_i x_{1+i} \cdots x_{j-1+i}(\alpha) \Big| 1 \leqslant j \leqslant r, 0 \leqslant i < \binom{n}{j} \right\}.$$

Then $B_2(r)$ is a basis of $\mathscr{S}_r$.

(ii) Let

$$x_0 x_1 \cdots x_{i-1}(\alpha) = \frac{g_i(x)}{G_i(x)} (1 \leqslant i \leqslant r), \quad \deg g_i(x) < \deg G_i(x)$$

and

$$\beta = \frac{h(x)}{G_r(x)}, \quad \deg h(x) < \deg G_r(x).$$

Then the equation

$$\phi(\alpha) = \beta$$

has a unique solution $\phi \in \Phi$ with $\deg \phi \leqslant r$. Moreover, ϕ can be computed by the following procedure:

(1) Take $h_r(x) = h(x)$

(2) Determine $\mu_r, h_{r-1}, \cdots, \mu_1, h_0$ successively by the following formula

$$\mu_j \equiv g_j^{-1} h_j (\bmod F_j)$$

$$\deg \mu_j < \deg F_j$$

$$h_{j-1} = [(g_j\mu_j)_{G_j} + h_j]/F_j.$$

Then $\phi = \sum_{j=1}^{r} \mu_j * x_0 x_1 \cdots x_{j-1}$.

Note Theorem 4.4(ii) gives another synthesis algorithm for the feedforward sequence $\phi(\alpha)$.

5. Proof of Theorem 4.1

First we have to prepare ground for this. In the following we shall sometimes discuss sequences over $\bar{\mathbb{F}}_2$. With sequences over $\bar{\mathbb{F}}_2$ we can also associate formal Laurent series in $\bar{\mathbb{F}}_2((x^{-1}))$ as their generating functions as we did for sequences over $\mathbb{F}_2$. All the operations for sequences over $\mathbb{F}_2$ can be carried over to sequences over $\bar{\mathbb{F}}_2$.

In particular, for $\lambda, \mu, \theta, \tau \in \bar{\mathbb{F}}_2^*$, we have

$$\frac{\lambda}{x-\theta} = \frac{\lambda x^{-1}}{1-\theta x^{-1}} = \sum_{i=0}^{\infty} \lambda \theta^i x^{-1-i} = (\lambda, \lambda\theta, \lambda\theta^2, \cdots, \lambda\theta^t, \cdots),$$

$$\frac{\mu}{x-\tau} = (\mu, \mu\tau, \mu\tau^2, \cdots, \mu\tau^t, \cdots).$$

We define a componentwise multiplication for sequences over $\bar{\mathbb{F}}_2$ as follows:

$$(a_0, a_1, \cdots, a_t, \cdots) \cdot (b_0, b_1, \cdots, b_t, \cdots) = (a_0 b_0, a_1 b_1, \cdots, a_t b_t, \cdots), \quad a_i, b_i \in \bar{\mathbb{F}}_2$$

Corresponding to this componentwise multiplication, we introduce the notation

$$\frac{\lambda}{x-\theta} \odot \frac{\mu}{x-\tau} = \sum_{i=0}^{\infty} \lambda\mu(\theta\tau)^i x^{-1-i}.$$

Then we have the following lemma.

Lemma 5.1 $\quad \dfrac{\lambda}{x-\theta} \odot \dfrac{\mu}{x-\tau} = \dfrac{\lambda\mu}{x-\theta\tau}.$

Furthermore, we need the following lemma.

Lemma 5.2 Let $f(x), g(x) \in \mathbb{F}_2[x]$, $f(x)$ be irreducible, $\deg f(x) = n$ and $\deg g(x) < \deg f(x)$. Let θ be a root of $f(x)$, then there exists an element $\lambda \in \mathbb{F}_2[\theta]$ such that

$$\frac{g(x)}{f(x)} = \sum_{i=0}^{n-1} \frac{\lambda^{2^i}}{(x-\theta^{2^i})}.$$

Proof   By partial fraction expansion there exist $\lambda_0, \lambda_1, \cdots, \lambda_{n-1} \in \mathbb{F}_2[\theta]$ such that
$$\frac{g(x)}{f(x)} = \sum_{i=0}^{n-1} \frac{\lambda_i}{(x - \theta^{2^i})}.$$
Put $\lambda = \lambda_0$, then by Galois theory and the uniqueness of the partial fraction expansion we have $\lambda_i = \lambda^{2^i}$. □

Lemma 5.3   Let
$$\alpha = \sum_{i=1}^{s} \frac{g_i}{f_i},$$
where $\deg g_i < \deg f_i$, then we have

(i) $f(\alpha, x) | [f_1, \cdots, f_s]$

(ii) If $(f_1, g_1) = 1$ and $(f_1, f_i) = 1$ for all $i \geqslant 2$, then
$$f_1(x) | f(\alpha, x).$$

Proof   (i) Set $f = [f_1, \cdots, f_s]$, then there exists $g \in \mathbb{F}_2[x]$ such that $\alpha = g/f$. By Lemma 2.8, $f(\alpha, x) | f$.

(ii) It is sufficient to prove the case $s = 2$. Now
$$\alpha = \frac{f_2 g_1 + f_1 g_2}{f_1 f_2}.$$
Notice that
$$(f_2 g_1 + f_1 g_2, f_1) = (f_2 g_1, f_1) = 1.$$
Set $h = (f_2 g_1 + f_1 g_2, f_2)$, then f_1/h. By Lemma 2.8,
$$f(x, \alpha) = \frac{f_1 f_2}{(f_1 f_2, f_2 g_1 + f_1 g_2)} = \frac{f_1 f_2}{(f_2, f_2 g_1 + f_1 g_2)} = \frac{f_1 f_2}{h}.$$
Thus, $hf(x, \alpha) = f_1 f_2$. But $(f_1, h) = 1$, therefore, $f_1 | f(x, \alpha)$. □

Now let us come to the proof of Theorem 4.1.

At first we prove that $\mathscr{S}_r \subset \Omega(G_r(x))$. It is sufficient to prove that $x_{l_1} \cdots x_{l_r}(\alpha) \in \Omega(G_r(x))$ for $0 \leqslant l_1 < \cdots < l_r < n$. By Lemma 5.2, we have
$$\alpha = \frac{g(x)}{f(x)} = \sum_{i=0}^{n-1} \frac{\lambda^{2^i}}{(x - \theta^{2^i})},$$
where θ is a root of $f(x)$ and $\lambda \in \mathbb{F}_2[\theta]$. By lemma 2.7, we have

$$x^l * \alpha = \sum_{i=0}^{n-1} x^l * \frac{\lambda^{2^i}}{(x - \theta^{2^i})} = \sum_{i=0}^{n-1} \frac{\lambda^{2^i} \theta^{2^i l}}{(x - \theta^{2^i})}$$

And then by Lemma 5.1,

$$x_{l_1} \cdots x_{l_r}(\alpha) = (x^{l_1} * \alpha) \cdots (x^{l_r} * \alpha)$$

$$= \sum_{0 \leq i_1 < n} \frac{\lambda^{2^{i_1}} \theta^{2^{i_1} l_1}}{x - \theta^{2^{i_1}}} \odot \cdots \odot \sum_{0 \leq i_r < n} \frac{\lambda^{2^{i_r}} \theta^{2^{i_r} l_r}}{x - \theta^{2^{i_r}}}$$

$$= \sum_{0 \leq i_1, \cdots, i_r < n} \frac{\lambda^{2^{i_1} + \cdots + 2^{i_r}} \theta^{2^{i_1} l_1 + \cdots + 2^{i_r} l_r}}{x - \theta^{2^{i_1} + \cdots + 2^{i_r}}} = \frac{H(x)}{G_r(x)},$$

where $H(x) \in \mathbb{F}_2[x]$ and $\deg H(x) < \deg G_r(x)$. Consequently, $x_{l_1} \cdots x_{l_r}(\alpha) \in \Omega(G_r(x))$.

Form $\mathscr{S}_r \subset \Omega(G_r(x))$ we deduce in particular $\mathscr{S}_n \subset \Omega(G_n(x))$. Clearly, distinct ϕ's in Φ give distinct feedforward sequences $\phi(\alpha)$'s. Consequently,

$$|\mathscr{S}_n| = |\Phi| = 2^{2^n - 1}.$$

But

$$\dim_{\mathbb{F}_2} \Omega(G_n(x)) = \deg G_n(x) = 2^n - 1.$$

thus

$$|\Omega(G_n(x))| = 2^{2^n - 1}.$$

It follows that $\mathscr{S}_n = \Omega(G_n(x))$. Furthermore, $B_1(n)$ linearly spans $\mathscr{S}_n$ over $\mathbb{F}_2$ and $|B_1(n)| = 2^n - 1$. Therefore, $B_1(n)$ is a basis of $\mathscr{S}_n$. However $B_1(r) \subset B_1(n)$ and $B_1(r)$ linearly spans $\mathscr{S}_r$ over $\mathbb{F}_2$, it follows that $B_1(r)$ is a basis of $\mathscr{S}_r$. Since

$$\dim_{\mathbb{F}_2} \mathscr{S}_r = |B_1(r)| = \sum_{i=0}^{r} \binom{n}{i} = \deg G_r(x) = \dim_{\mathbb{F}_2} \Omega(G_r(x))$$

and $\mathscr{S}_r \subset \Omega(G_r(x))$, we have $\mathscr{S}_r = \Omega(G_r(x))$.

Theorem 4.1 is completely proved. □

6. Proofs of Theorems 4.2 and 4.3

Proof of Theorem 4.2. Following the notation of the proof of Theorem 4.1, we have

$$x_{l_1}\cdots x_{l_r}(\alpha) = \sum_{0\leqslant i_1,\cdots,i_r<n} \frac{\lambda^{2^{i_1}+\cdots+2^{i_r}}\theta^{2^{i_1}l_1+\cdots+2^{i_r}l_r}}{x-\theta^{2^{i_1}+\cdots+2^{i_r}}}.$$

Set

$$\sigma_r = \sum_{\substack{0\leqslant i_1,\cdots,i_r<n \\ i_j\neq i_k,\text{if }j\neq k}} \frac{\lambda^{2^{i_1}+\cdots+2^{i_r}}\theta^{2^{i_1}l_1+\cdots+2^{i_r}l_r}}{x-\theta^{2^{i_1}+\cdots+2^{i_r}}}.$$

and

$$\beta = \sum_{\substack{0\leqslant i_1,\cdots,i_r<n \\ i_j=i_k,\text{for some }j\neq k}} \frac{\lambda^{2^{i_1}+\cdots+2^{i_r}}\theta^{2^{i_1}l_1+\cdots+2^{i_r}l_r}}{x-\theta^{2^{i_1}+\cdots+2^{i_r}}}.$$

Then

$$x_{l_1}\cdots x_{l_r}(\alpha) = \sigma_r + \beta.$$

We may write

$$\sigma_r = \sum_{\substack{0\leqslant t<2^n-1 \\ w(t)=r}} \frac{\lambda^t}{x-\theta^t} \sum_{\substack{\binom{i_1\cdots i_r}{t_1\cdots t_r}\in S_r \\ t=2^{t_1}+\cdots+2^{t_r} \\ t_1<\cdots<t_r}} \theta^{2^{i_1}l_1+\cdots+2^{i_r}l_r}$$

$$= \sum_{\substack{0\leqslant t<2^n-1 \\ w(t)=r}} \frac{\lambda^t}{x-\theta^t} \begin{vmatrix} \theta^{2^{t_1}l_1} & \cdots & \theta^{2^{t_1}l_r} \\ \vdots & & \vdots \\ \theta^{2^{t_r}l_1} & \cdots & \theta^{2^{t_r}l_r} \end{vmatrix},$$

where S_r is the symmetric group of r letters $t_1,\cdots,t_r$. Put

$$\delta_\tau(l_1,\cdots,l_r) = \begin{vmatrix} \theta^{2^{t_1}l_1} & \cdots & \theta^{2^{t_1}l_r} \\ \vdots & & \vdots \\ \theta^{2^{t_r}l_1} & \cdots & \theta^{2^{t_r}l_r} \end{vmatrix}.$$

Clearly, $\beta \in \Omega(G_{r-1}(x))$. If $\delta_t^{(l_1,\cdots,l_r)} \neq 0$, then $x-\theta^t | f(x_{l_1\cdots l_r}(\alpha),x)$. Therefore, if $\delta_t^{(l_1,\cdots,l_r)} \neq 0$ for all t such that $0\leqslant t<2^n-1$ and $w(t)=r$, then $F_r(x)|f(x_{l_1}\cdots x_{l_r}(\alpha),x)$. Now for $x_0 x_l \cdots x_{(r-1)l}$, we have

$$\delta_t^{(0,1,\cdots,r-1)} = \begin{vmatrix} 1 & \theta^{2^{t_1}l} & \cdots & \theta^{2^{t_1}(r-1)l} \\ \vdots & \vdots & & \vdots \\ 1 & \theta^{2^{t_r}l} & \cdots & \theta^{2^{t_r}(r-1)l} \end{vmatrix} \neq 0,$$

provided that $\theta^{2^{t_r l}}, \cdots, \theta^{2^{t_r l}}$ are pairwise distinct. If $\theta^{2^{t_r l}} = \theta^{2^{t_r l}}$, $i < j$, then putting $d = t_j - t_i$ we have $\theta^{2^{t_r l(2^d - 1)}} = 1$, hence $2^{t_i} l(2^d - 1) = 0 (\mathrm{mod}(2^n - 1))$ and hence $l \equiv 0(\mathrm{mod}(2^n - 1)/2^\tau - 1)$, where $\tau = (d, n)$. This contradicts the assumption of Theorem 4.2. □

In order to prove Theorem 4.3 we need some preparation.

Lemma 6.1 $x^k * (x^i * \phi(x_0, \cdots, x_{n-1})) \equiv x^{k+i} * \phi(x_0, \cdots, x_{n-1})$.

The proof is easy and will be omitted.

Lemma 6.2 Let $\phi(x_0, \cdots, x_{n-1}) \in \Phi$ and $\mu(x) \in \mathbb{F}_2[x]$. Then

$$(\mu(x) * \phi(x_0, \cdots, x_{n-1}))(\alpha) = \mu(x) * (\phi(x_0, \cdots, x_{n-1})(\alpha)).$$

Proof Let $\mu_1, \mu_2 \in \mathbb{F}_2[x]$, and $\phi_1, \phi_2 \in \Phi$, then

$$((\mu_1 + \mu_2) * (\phi_1 + \phi_2))(\alpha) = \sum_{i,j=1}^{2} (\mu_i * \phi_j)(\alpha),$$

$$((\mu_1 + \mu_2) * (\phi_1 + \phi_2))(\alpha) = \sum_{i,j=1}^{2} \mu_i * (\phi_j(\alpha)).$$

Thus, it is sufficient to prove that

$$(x^s * x_{l_1} \cdots x_{l_r})(\alpha) = x^s * (x_{l_1} \cdots x_{l_r}(\alpha)).$$

Since $x^i \equiv \sum a_{ij} x^j (\mathrm{mod} f(x))$, we have

$$\left(x^i - \sum a_{ij} x^j\right)(\alpha) = 0$$

and

$$a_{i+t} = \sum_{0 \leqslant j < n} a_{ij} a_{j+t}, \quad i \geqslant 0.$$

Let

$$(x^s * x_{l_1} \cdots x_{l_r})(\alpha) = (b_0, b_1, \cdots, b_t, \cdots),$$

$$(x^s * x_{l_1} \cdots x_{l_r}(\alpha)) = (c_0, c_1, \cdots, c_t, \cdots),$$

then

$$b_t = \left(\left(\sum_{0 \leqslant j < n} a_{l_1 + s_j} x_j\right) \cdots \left(\sum_{0 \leqslant j < n} a_{l_r + s_j} x_j\right)\right)(a_t, a_{t+1}, \cdots, a_{t+n-1})$$

$$= \left(\sum a_{x_1+s_j} a_{t+j}\right) \cdots \left(\sum a_{l_r+s_j} a_{t+j}\right)$$
$$= a_{l_1+s+t} a_{l_2+s+t} \cdots a_{l_r+s+t} = c_t.$$

This proves Lemma 6.2. □

Lemma 6.3 Let $\rho \in \mathbb{F}_2(\theta)^*$ and suppose $\deg f(\rho, x) = d$, then $(n/d)|w(\rho)$.

Proof By the theory of finite fields, ρ belongs to the cyclic group generated by $\theta^{(2^n-1)/(2^d-1)}$ and thus there exists an integer c, $0 \leqslant c < 2^d - 1$ such that
$$\rho = \theta^{(2^n-1)c/(2^d-1)}.$$

Express c in dyadic system, $c = \sum_{i=1}^{r} 2^{c_i}$, $0 \leqslant c_1 < c_2 < \cdots < c_r < d$. Then obviously
$$w(\rho) = w\left(\frac{2^{n-1}}{2^{d-1}} \cdot c\right) \frac{n}{d} \cdot r.$$

Therefore, $(n/d)|w(\rho)$. □

Corollary 6.4 Let $\rho \in \mathbb{F}_2(\theta)^*$ and $w(\rho) = r$, where $1 \leqslant r \leqslant n$ and $(r, n) = 1$. Then $\deg f(\rho, x) = n$.

Proof Let $\deg f(\rho, x) = d$. By Lemma 6.3, $(n/d)|r$. Thus $(n/d)|(r, n)$. But $(r, n) = 1$, hence $n = d$. □

Proof of Theorem 4.3 (i) By hypothesis and Lemma 5.1, we have $\phi(\alpha) = \beta + \gamma$, where $f(\beta, x) = F_r(x), f(\gamma, x)|G_{r-1}(x)$. Thus, we can assume $\beta = g(x)/F_r(x)$, $(g(x), F_r(x)) = 1$. By hypothesis $(r, n) = 1$, thus from the Corollary 6.4 we deduce that the irreducible factors of $F_r(x)$ are all of degree n. But $\deg \mu(x) < n$, hence $(\mu(x), F_r(x)) = 1$ and
$$((\mu(x)g(x))_{F_r(x)}, F_r(x)) = 1.$$

By Lemma 2.7
$$\mu(x) * \beta = \frac{(\mu(x)g(x))_{F_r(x)}}{F_r(x)}$$
then by Lemma 2.8, $f(\mu(x) * \beta, x) = F_r(x)$. By Lemma 6.3
$$(\mu * \phi + \psi)(\alpha) = \mu * \phi(\alpha) + \psi(\alpha) = \mu * \beta + \mu * \gamma + \psi(\alpha).$$

But
$$f((\mu * \gamma + \psi)(\alpha), x) | G_{r-1}(x).$$

Thus by Lemma 5.2
$$F_r(x) | f((\mu * \phi + \psi)(\alpha), x).$$

By hypothesis $\deg \phi = r$, $\deg \psi < r$, thus $\deg(\mu * \phi + \psi) = r$. Therefore, $\mu * \phi + \psi$ is a nondegenerate feedforward transformation of degree r. This proves (i).

(ii) If $\mu * \phi + \psi = \mu_1 * \phi + \psi_1$, then $(\mu - \mu_1) * \phi + (\psi - \psi_1) = 0$. By (i), $\mu = \mu_1$, and hence $\psi = \psi_1$. This proves (ii). □

7. Proof of Theorem 4.4

With the preparations carried out, we can now take up to the proof of Theorem 4.4.

Proof of Theorem 4.4 (i) is a consequence of the assertion that $\phi = \sum_{j=1}^{r} \mu_j * x_c x_1 \cdots x_{j-1}$ is a solution of $\phi(\alpha) = \beta$ in (ii). In fact, write
$$\mu_j = \sum_{0 \leqslant i < \binom{n}{j}} c_{ji} x^i.$$

Then by Lemma 6.2
$$(\mu_j * x_0 x_i \cdots x_{j-1})(\alpha) = \mu_j * ((x_0 x_1 \cdots x_{j-1})(\alpha))$$
$$= \left(\sum c_{ji} x^i \right) * ((x_0 x_1 \cdots x_{j-1})(\alpha))$$
$$= \sum_i (c_{ji} x_i x_{i+1} \cdots x_{i+j-1})(\alpha).$$

Consequently,
$$\phi(\alpha) = \sum_{1 \leqslant j < r} \sum_{0 \leqslant i < \binom{n}{j}} (c_{ji} x_i x_{i+1} \cdots x_{i+j-1})(\alpha),$$

i.e. $\mathscr{S}_r$ can be linearly spanned by $B_2(r)$ over $\mathbb{F}_2$. Since
$$|B_2(r)| = \sum_{j=1}^{r} \binom{n}{j} = \dim_{\mathbb{F}_2} \mathscr{S}_r.$$

$B_2(r)$ is a basis of $\mathscr{S}_r$.

(ii) The procedure of the computation is effective. In fact, by Theorem 4.2, $x_0 x_1 \cdots x_{j-1}$ is a nondegenerate feedforward transformation of degree r, hence $(F_j(x), g_j(x)) = 1$, and hence $g_j^{-1} \pmod{F_j}$ exists. By the formulas of the computation, $\deg h_j < \deg G_j$.

We have $G_r = F_r G_{r-1}$, $1 \leqslant r \leqslant n$ (we agree that $G_0(x) = 1$). Now we assert that

$$\sum_{i=j+1}^{r} \mu_i * \frac{g_i}{G_i} + \frac{h}{G_r} = \frac{h_j}{G_j}, \quad 0 \leqslant j \leqslant r, \tag{2}$$

$$h_0(x) = 0. \tag{3}$$

We apply backward induction to prove (2). For $j = r$, (2) holds obviously. Assume that (2) holds for $j \geqslant k$, we are going to prove that (2) also holds for $j = k - 1$. By definition

$$\mu_r g_k = h_k (\bmod F_k),$$

$$((\mu_k g_k)_{G_k} + h_k)/F_k = h_{k-1}.$$

But

$$\mu_k * \frac{g_k}{G_k} = \frac{(\mu_k g_k)_{G_k}}{G_k} = \frac{F_k h_{k-1} + h_k}{G_k} = \frac{h_{k-1}}{G_k} + \frac{h_k}{G_k},$$

thus

$$\sum_{i=k}^{r} \mu_i * \frac{g_i}{G_i} + \frac{h}{G_r} = \sum_{i=k+1}^{r} \mu_i * \frac{g_i}{G_i} + \mu_k * \frac{g_k}{G_k} + \frac{h}{G_r}$$

$$= \frac{h_k}{G_k} + \mu_k * \frac{g_k}{G_k} \text{ (by induction hypothesis)}$$

$$= \frac{h_{k-1}}{G_{k-1}}.$$

This proves (2).

We know that $F_1 = G_1$. Then by definition,

$$\mu_1 g_1 \equiv h_1 (\bmod F_1),$$

$$(\mu_1 g_1)_{G_1} = (\mu_1 g_1)_{F_1} \equiv h_1 (\bmod F_1),$$

$$(\mu_1 g_1)_{G_1} + h_1 \equiv 0 \;(\text{mod } F_1).$$

But

$$\deg((\mu_1 g_1)_{G_1} + h_1) < \deg G_1 = \deg F_1,$$

hence $(\mu_1 g_1)_{G_1} + h_1 = 0$, and consequently, $h_0 = 0$. Take $j = 0$ in (2), we have

$$\sum_{1}^{r} \mu_i * \frac{g_i}{G_i} + \frac{h}{G_r} = 0.$$

Consequently,

$$\beta = \frac{h}{G_r} = \sum \mu_i * x_0 x_1 \cdots x_{i-1}(\alpha)$$
$$= \sum (\mu_i * x_0 x_1 \cdots x_{i-1})(\alpha).$$

Therefore, $\phi = \sum_{i=1}^{r} \mu_i * x_0 x_1 \cdots x_{i-1}$ is a solution of the equation $\phi(\alpha) = \beta$. From this fact and (i) we conclude that ϕ is also the unique solution. $\square$

References

[1] E.J. Groth, Generation of binary sequences with controllable complexity, IEEE Trans. Inform. Theory 17 (1971) 288–296.

[2] E.L. Key, An analysis of the structure and complexity of nonlinear binary sequence generators. IEEE Trans. Inform. Theory 22 (1976) 732–736.

A Linear Algebra Approach to Minimal Convolutional Encoders

Rolf Johannesson, *Member, IEEE*, and Zhexian Wan

Abstract This semitutorial paper starts with a review of some of Forney's contributions on the algebraic structure of convolutional encoders on which some new results on minimal convolutional encoders rest. An example is given of a basic convolutional encoding matrix whose number of abstract states is minimal over all equivalent encoding matrices. However, this encoding matrix can be realized with a minimal number of memory elements neither in controller canonical form nor in observer canonical form. Thus, this encoding matrix is not minimal according to Forney's definition of a minimal encoder. To resolve this difficulty, the following three minimality criteria are introduced: *minimal-basic encoding matrix* (minimal overall constraint length over equivalent basic encoding matrices), *minimal encoding matrix* (minimal number of abstract states over equivalent encoding matrices), and *minimal encoder* (realization of a minimal encoding matrix with a minimal number of memory elements over all realizations). Among other results, it is shown that all minimalbasic encoding matrices are minimal, but that there exist (basic) minimal encoding matrices that are not minimal-basic! Several equivalent conditions are given for an encoding matrix to be minimal. It is also proven that the constraint lengths of two equivalent minimal-basic

---

Manuscript received June 19,1991; revised November17,1992. This work was supported in part by the Swedish Research Council for Engineering Sciences under Grant 91.

R. Johannesson is with the Department of Information Theory, University of Lund, Box 118, S-221 00,Sweden.

Z.-x. Wan is with the Department of Information Theory, University of Lund, Box 118, S-221 00, Sweden. He is also with the Institute for Systems Science, Chinese Academy of Sciences, Beijing 1000 80, China.

IEEE Log Number 9209457.

encoding matrices are equal one by one up to a rearrangement. All results are proven using only elementary linear algebra. Most important among the new results are a simple minimality test, the surprising fact that there exist basic encoding matrices that are minimal but not minimal-basic, the existence of basic encoding matrices that are nonminimal, and a recent result, due to Forney, that states exactly when a basic encoding matrix is minimal.

Index Terms Convolutional code, basic encoding matrix, minimal-basic encoding matrix, minimal encoding matrix, minimal encoder.

I. Introduction

Forney's landmark paper: "Convolutional codes I: Algebraic structures"[1] (see also [2]) constitutes an abundant source of important results on the algebraic structure of convolutional codes. After having introduced a most important concept, viz., a *basic* convolutional encoder (which has both a polynomial encoding matrix and a polynomial right inverse), Forney defined a rate $R = b/c$ basic encoder to be *minimal* if its overall constraint length ν is equal to the maximum degree μ of the $b \times b$ subdeterminants of its encoding matrix.

In this semitutorial paper, we show that there exist basic encoding matrices that have a minimal number of abstract states and, hence, can be realized by a minimal number of memory elements over all equivalent encoders but are, quite surprisingly, *not* minimal according to Forney's definition! To resolve this difficulty we introduce the following three minimality criteria: *minimal-basic encoding matrix* (minimal overall constraint length over equivalent basic encoding matrices), *minimal encoding matrix* (minimal number of abstract states over equivalent encoding matrices), and *minimal encoder* (realization of a minimal encoding matrix with a minimal number of memory elements over all realizations). Our definition of a minimal-basic encoding matrix is equivalent to Forney's definition of a minimal encoder.

In Section II, we introduce the controller canonical and observer canonical forms of a linear sequential circuit. The distinction between convolutional encoders and their generator and encoding matrices is discussed in Section III.

In Section IV, we discuss the equivalence of encoders and we also give Forney's definition of a basic encoder. The important concept of minimal-basic encoding matrices is introduced in Section V. In this section, we give three equivalent statements for a basic encoding matrix to be minimal-basic. We also prove that the constraint lengths of two equivalent minimal-basic encoding matrices are equal one by one up to a rearrangement. From the minimal-basic encoding matrix we proceed to introduce the minimality of a general encoding matrix in Section VI. Several equivalent conditions for a general encoding matrix to be minimal are given. After having defined a minimal encoder in Section VII, we give an example of a basic minimal encoding matrix that is not minimal-basic and, hence, has no minimal realization in controller canonical form. Finally, in Section VIII, we prove that every systematic encoding matrix is minimal. We also show a minimal realization of a systematic encoding matrix which has neither a minimal realization in controller canonical form nor one in observer canonical form.

Some of our theorems are new, others can be found explicitly in Forney's papers [1]-[3], and a few are given implicitly in these papers, but we have proven all results by using only elementary linear algebra. Most important among the new results are a simple minimality test, the surprising fact that there exist basic encoding matrices that are minimal but not minimal-basic, the existence of basic encoding matrices that are nonminimal, and a recent result, due to Forney, that states exactly when a basic encoding matrix is minimal.

II. Controller and Observer Canonical Forms

Let $\boldsymbol{F}_2((D))$ denote the *field of binary Laurent series*. The element $x(D) = \sum_{i=r}^{\infty} x_i D^i \in \boldsymbol{F}_2((D)), r \in \boldsymbol{Z}$, contains only finitely many negative powers of D. For example,

$$x(D) = D^{-2} + 1 + D^3 + D^7 + D^{12} + \cdots$$

is a Laurent series.

Let $\boldsymbol{F}_2[[D]]$ denote the *ring of formal power series*. The element $f(D) = \sum_{i=0}^{\infty} f_i D^i \in \boldsymbol{F}_2[[D]]$ is a Laurent series without negative powers of D.

A *polynomial* $p(D) = \sum_{i=0}^{\infty} p^i D^i$ contains no negative and only finitely many positive powers of D. If $p_0 = 1$ we have a *delayfree* polynomial, e.g.,

$$p(D) = 1 + D^2 + D^3 + D^5$$

is a binary delayfree polynomial of degree 5. The set of binary polynomials $\boldsymbol{F}_2[D]$ is clearly a subset of $\boldsymbol{F}_2((D))$.

Given any pair of polynomials $x(D), y(D) \in \boldsymbol{F}_2[D]$, with $y(D) \neq 0$, we can obtain the element $x(D)/y(D) \in \boldsymbol{F}_2((D))$ by long division. Since sequences must start at some finite time we must identify, for instance, $(1+D)/D^2(1+D+D^2)$ with the series $D^{-2} + 1 + D + D^3 + \cdots$ instead of the alternative series $D^{-3} + D^{-5} + D^{-6} + \cdots$ that can also be obtained by long division but which is not a Laurent series. Obviously, all nonzero ratios $x(D)/y(D)$ are invertible, so they form the *field of binary rational functions* $\boldsymbol{F}_2((D))$, which is a subfield of the field of Laurent series $\boldsymbol{F}_2((D))$.

We can of course consider n-tuples of elements from $\boldsymbol{F}_2[D], \boldsymbol{F}_2[[D]], \boldsymbol{F}_2(D)$, or $\boldsymbol{F}_2((D))$. For example, $x(D) = \left(\sum_{i=r_1}^{\infty} x_i^{(1)} D^i, \sum_{i=r_2}^{\infty} x_i^{(2)} D^i, \cdots, \sum_{i=r_n}^{\infty} x_i^{(n)} D^i \right)$, where $r_1, r_2, \cdots, r_n \in \boldsymbol{Z}$, is an element in $\boldsymbol{F}_2((D))^{(n)}$, the n-dimensional vector space over the field of binary Laurent series $\boldsymbol{F}_2((D))$. Let $r = \min\{r_1, r_2, \cdots, r_n\}$ and put $x_i^{(j)} = 0$ for $i < r_j$, then we can express $x(D)$ also as

$$x(D) = \sum_{i=r}^{\infty} (x_i^{(1)}, x_i^{(2)}, \cdots, x_i^{(n)}) D^i.$$

Thus, $\boldsymbol{F}_2((D))^{(n)} = \boldsymbol{F}_2^{(n)}((D))$, where $\boldsymbol{F}_2^{(n)}((D))$ is the set of all Laurent series in D with coefficients in the n-dimensional row vector space $\boldsymbol{F}_2^{(n)}$ over $\boldsymbol{F}_2$. The elements in the field $\boldsymbol{F}_2((D))$ are usually called scalars. Similarly, $\boldsymbol{F}_2[D]^{(n)} = \boldsymbol{F}_2^{(n)}[D]$, $\boldsymbol{F}_2[[D]]^{(n)} = \boldsymbol{F}_2^{(n)}[[D]]$, etc. If $x(D) \in \boldsymbol{F}_2^{(n)}[D]$, we say that $\boldsymbol{x}(D)$ is polynomial in D. The degree of the element $x(D) = \sum_{i=0}^{m}(x_i^{(1)}x_i^{(2)}\cdots x_i^{(n)})D^i$ is defined to be m, provided $(x_m^{(1)}x_m^{(2)}\cdots x_m^{(n)}) \neq (00\cdots 0)$. For simplicity, we also call elements in $\boldsymbol{F}_2^{(n)}[[D]]$ formal power series when $n > 1$.

Consider the *controller canonical form* of a single input single output linear system as shown in Fig. 1. The delay elements form a shift register, the output is a linear function of the input and the shift register contents, and the input to the shift register is a linear function of the input and the shift register contents.

From Fig. 1, it follows that

$$v(D) = u(D)g(D)/q(D), \tag{1}$$

where

$$g(D) = g_0 + g_1 D + \cdots + g_m D^m \tag{2}$$

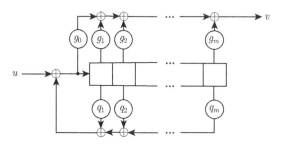

Fig. 1 The controller canonical form of a rational transfer function

and

$$q(D) = 1 + q_1 D + \cdots + q_m D^m. \tag{3}$$

Let $t(D) = g(D)/q(D)$, then $v(D) = u(D)t(D)$ and we say that $t(D)$ is a *rational transfer function* which transfers the input $u(D)$ into the output $v(D)$. From (1), it follows that every rational function with a constant term 1 in the denominator polynomial $q(D)$ (or, equivalently, with $q(0) = 1$ or, again equivalently, with $q(D)$ delayfree) is a rational transfer function that can be realized in the canonical form shown in Fig. 1. Every rational function $g(D)/q(D)$, where $q(D)$ is delayfree, is called a *realizable function*.

In general, a matrix $G(D)$ whose entries are rational functions is called a *rational transfer function matrix*. A rational transfer function matrix $G(D)$ for a linear system with many inputs and/or many outputs whose entries are realizable functions is called *realizable*.

In practice, given a rational transfer function matrix we have to realize it by linear sequential circuits. It can be realized in many different ways. For instance, the realizable function

$$\frac{g_0 + g_1 D + \cdots + g_m D^m}{1 + q_1 D + \cdots + q_m D^m} \tag{4}$$

has the controller canonical form illustrated in Fig. 1. On the other hand, since the circuit in Fig. 2 is linear, we have

$$\begin{aligned}v(D) =& u(D)(g_0 + g_1 D + \cdots + g_m D^m) \\ &+ v(D)(q_1 D + \cdots + q_m D^m),\end{aligned} \tag{5}$$

which is the same as (1). Thus, Fig. 2 is also a realization of (4). In this realization the delay elements do not in general form a shift register as these delay elements are separated by adders. This is the so-called *observer canonical form* of the rational function (4). The controller and observer canonical forms in Fig. 1 and 2, respectively, are two different realizations of the same rational transfer function.

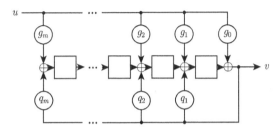

Fig. 2   The observer canonical form of a rational transfer function

III. Convolutional Codes and Their Encoders

We are now prepared to give a formal definition of a convolutional transducer.

Definition A rate $R = b/c$ (binary) *convolutional transducer* over the field of rational functions $\boldsymbol{F}_2(D)$ is a linear mapping

$$\boldsymbol{F}_2^b((D)) \to \boldsymbol{F}_2^c((D))$$
$$\boldsymbol{u}((D)) \mapsto \boldsymbol{v}(D), \tag{6}$$

which can be represented as

$$\boldsymbol{v}(D) = \boldsymbol{u}(D)G(D), \tag{7}$$

where $G(D)$ is a $b \times c$ transfer function matrix of rank b with entries in $\boldsymbol{F}_2(D)$ and $\boldsymbol{v}(D)$ is called a *code sequence* arising from the *information sequence* $\boldsymbol{u}(D)$.

Obviously we must be able to reconstruct the information sequence $\boldsymbol{u}(D)$ from the code sequence $\boldsymbol{v}(D)$ when there is no noise on the channel. Otherwise the convolutional transducer would be useless. Therefore we require that the transducer map is injective, i.e., the transfer function matrix $G(D)$ has rank b over the field $\boldsymbol{F}_2(D)$.

Definition A rate $R = b/c$ *convolutional code* C over $\boldsymbol{F}_2$ with the $b \times c$ matrix $G(D)$ of rank b over $\boldsymbol{F}_2(D)$ as its *generator matrix* is the image set of a rate $R = b/c$ convolutional transducer with $G(D)$ as its transfer function matrix.

It follows immediately from the definition that a rate $R = b/c$ convolutional code C over $\boldsymbol{F}_2$ with the $b \times c$ matrix $G(D)$ of rank b over $\boldsymbol{F}_2(D)$ as a generator matrix can be regarded as the $\boldsymbol{F}_2((D))$ row space of $G(D)$. Hence, it can also be regarded as the rate $R = b/c$ block code over the infinite field of Laurent series which has $G(D)$ as its generator matrix.

We call a realizable transfer function matrix $G(D)$ *delayfree* if at least one of its entries $g(D)/q(D)$ has $g(0) \neq 0$. If $G(D)$ is not delayfree it can be written as

$$G(D) = D^i G_d(D), \qquad (8)$$

where $i \geqslant 1$ and $G_d(D)$ is delayfree.

Definition The generator matrix of a convolutional code over $\boldsymbol{F}_2$ is called an *encoding matrix* of the code if it is (realizable and) delayfree.

Definition A rate $R = b/c$ *convolutional encoder* of a convolutional code with encoding matrix $G(D)$ over $\boldsymbol{F}_2(D)$ is a realization by a linear sequential circuit of a rate $R = b/c$ convolutional transducer whose transfer function matrix is $G(D)$.

Theorem 1 *Every convolutional code C has an encoding matrix.*

Proof Let $G(D)$ be any generator matrix for C. The nonzero entries of $G(D)$ can be written
$$D^{s_{ij}} g_{ij}(D)/q_{ij}(D), \qquad (9)$$
where s_{ij} is an integer such that $g_{ij}(0) = q_{ij}(0) = 1, 1 \leqslant i \leqslant b, 1 \leqslant j \leqslant c$. The number s_{ij} is called the *start* of the sequence
$$t(D) = D^{s_{ij}} g_{ij}(D)/q_{ij}(D) = D^{s_{ij}} + t_{s_{ij}+1} D^{s_{ij}+1} + \cdots. \qquad (10)$$

Let $s = \min_{i,j}\{s_{ij}\}$. Clearly
$$G'(D) = D^{-s} G(D) \qquad (11)$$

is both delayfree and realizable. Since D^{-s} is a scalar in $\boldsymbol{F}_2((D))$, both $G(D)$ and $G'(D)$ generate the same convolutional code. Therefore $G'(D)$ is an encoding matrix for the convolutional code C. □

Including nondelayfree generator matrices in the set of convolutional encoders would only clutter up the analysis without any benefits.

A given convolutional code can be encoded by many essentially different encoders. For example, consider the rate $R = 1/2$, binary convolutional code with the basis vector $v_0(D) = (1 + D + D^2 \quad 1 + D^2)$. The simplest encoder for this code has the encoding matrix
$$G_0(D) = (1 + D + D^2 \quad 1 + D^2). \qquad (12)$$

Its controller canonical form is shown in Fig. 3.

If we choose the basis to be $\boldsymbol{v}_1(D) = a_1(D)\boldsymbol{v}_0(D)$, where the scalar $a_1(D)$ is the rational function $a_1(D) = 1/(1+D+D^2)$, we obtain the encoding matrix

$$G_1(D) = \begin{pmatrix} 1 & \dfrac{1+D^2}{1+D+D^2} \end{pmatrix}, \tag{13}$$

which is a *systematic* encoding matrix for the same code (Fig. 4). When a rate $R = b/c$ convolutional code is encoded by a systematic encoding matrix, the b input sequences appear unchanged among the c output sequences.

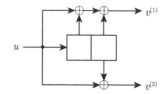

Fig. 3 A rate $R = 1/2$ convolutional encoder with encoding matrix $G_0(D)$

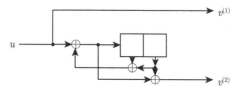

Fig. 4 A rate $R = 1/2$ systematic convolutional encoder with feedback and encoding matrix $G_1(D)$

For example, the output sequence $\boldsymbol{v}(D) = (v^{(1)}(D) v^{(2)}(D))$ of the systematic convolutional encoder with encoding matrix $G_1(D)$ shown in Fig. 4 can be written as

$$\begin{aligned} v^{(1)}(D) &= u(D) \\ v^{(2)}(D) &= u(D)\dfrac{1+D^2}{1+D+D^2}. \end{aligned} \tag{14}$$

If a convolutional code C is encoded by a systematic encoding matrix we can always permute its columns and obtain an encoding matrix for an *equivalent* convolutional code C' such that the b information sequences appear unchanged among the *first* c code sequences. Thus, without loss of generality, a systematic encoding matrix can be written

$$G(D) = (I_b R(D)), \qquad (15)$$

where I_b is a $b \times b$ identity matrix and $R(D)$ a $b \times (c-b)$ matrix whose entries are rational functions of D.

Being "systematic" is an encoding matrix property, not a code property. Every convolutional code has both systematic and *nonsystematic* encoding matrices.

If we further change the basis to $v_2(D) = a_2(D)v_0(D)$, where $a_2(D) \in F_2(D)$ is chosen as $a_2(D) = 1 + D$, we obtain a third encoding matrix for the same code, viz.,

$$G_2(D) = (1 + D^3 \quad 1 + D + D^2 + D^3). \qquad (16)$$

In Fig. 5, we show its controller canonical form.

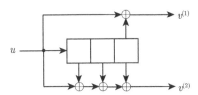

Fig. 5   A rate $R = 1/2$ catastrophic convolutional encoder with encoding matrix $G_2(D)$

Theorem 2   *Every convolutional code C has a polynomial encoding matrix.*

Proof   Let $G(D)$ be any encoding matrix for C and let $q(D)$ be the least common multiple of all the denominators in (9). Since $q(D)$ is a delayfree scalar in $F_2((D))$,

$$G'(D) = q(D)G(D) \qquad (17)$$

is a polynomial encoding matrix for C. □

An encoder whose encoding matrix is polynomial is called a polynomial encoder.

For convolutional codes, the choice of the encoding matrix is of great importance.

Definition A generator matrix for a convolutional code is *catastrophic*[1] [4] if there exists an information sequence $u(D)$ with infinitely many nonzero digits, $w_H(u(D)) = \infty$, that results in codewords $v(D)$ with only finitely many nonzero digits, $w_H(v(D)) < \infty$.

Example 1 The third encoding matrix for the convolutional code given above, viz.,

$$G_2(D) = (1 + D^3 \quad 1 + D + D^2 + D^3)$$

is catastrophic since $u(D) = 1/(1+D) = 1 + D + D^2 + \cdots$ has $w_H(u(D)) = \infty$ but $v(D) = u(D)G_2(D) = (1+D+D^2 \quad 1+D^2) = (1 \quad 1) + (1 \quad 0)D + (1 \quad 1)D^2$ has $w_H(v(D)) = 5 < \infty$.

When a catastrophic encoding matrix is used for encoding, finitely many errors (five in the previous example) in the estimate $\hat{v}(D)$ of the transmitted codeword $v(D)$ can lead to infinitely many errors in the estimate $\hat{u}(D)$ of the information sequence $u(D)$—a "catastrophic" situation that must be avoided!

Being "catastrophic" is a generator matrix property, not a code property. Every convolutional code has both catastrophic and noncatastrophic generator matrices.

We define the *constraint length for the ith input* of a polynomial convolutional encoding matrix as [1]

$$\nu_i = \max_{1 \leqslant j \leqslant c} \{\deg g_{ij}(D)\}, \tag{18}$$

the *memory* m of the polynomial encoding matrix as the maximum of the constraint lengths, i.e.,

$$m = \max_{1 \leqslant i \leqslant b} \{\nu_i\}, \tag{19}$$

---

[1] The term "catastrophic" does not actually appear in [4]; it seems to have been introduced by Massey in a seminar in 1969.

and the *overall constraint length* as the sum of the constraint lengths [1]

$$\nu = \sum_{i=1}^{b} \nu_i. \qquad (20)$$

The polynomial encoding matrix can be realized by a linear sequential circuit consisting of b shift registers, the ith of length ν_i, with the outputs formed as modulo-2 sums of the appropriate shift register contents. For example, in Fig. 6 we have shown the controller canonical form of the polynomial encoding matrix

$$G(D) = \begin{pmatrix} 1+D & D & 1 \\ D^2 & 1 & 1+D+D^2 \end{pmatrix}, \qquad (21)$$

whose constraint lengths of the 1st and 2nd inputs are 1 and 2, respectively, and whose overall constraint length is 3.

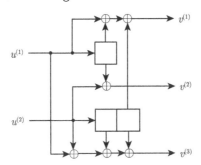

Fig. 6 A rate $R = 2/3$ convolutional encoder

The number of memory elements required for the controller canonical form is equal to the overall constraint length.

We define a *physical state* σ of a realization of a rational encoding matrix $G(D)$ at some time instant to be the contents of its memory elements. If $G(D)$ is polynomial, then the dimension of the physical state space of its controller canonical form is equal to the overall constraint length ν.

For a rational encoding matrix $G(D)$ we define the *abstract state* $s(D)$ associated with an input sequence $u(D)$ to be the sequence of outputs at time 0 and later, which are due to that part of $u(D)$ that occurs up to time -1, and to the all zero inputs thereafter. The abstract state depends only on $G(D)$ and not on its realization. Distinct abstract states must spring from distinct physical states at time 0. Clearly, the number of physical states is greater than or equal to the number of abstract states.

Let P be the projection operator that truncates sequences to end at time -1, and $Q = 1 - P$ the projection operator that truncates sequences to start at time 0, i.e., if
$$\boldsymbol{u}(D) = \boldsymbol{u}_d D^d + \boldsymbol{u}_{d+1} D^{d+1} + \cdots,$$
then
$$\boldsymbol{u}(D)P = \begin{cases} \boldsymbol{u}_d D^d + \cdots + \boldsymbol{u}_{-1} D^{-1}, & d < 0, \\ \boldsymbol{0}, & d \geqslant 0, \end{cases} \tag{22}$$
and
$$\boldsymbol{u}(D)Q = \boldsymbol{u}_0 + \boldsymbol{u}_1 D + \boldsymbol{u}_2 D^2 + \cdots. \tag{23}$$
Clearly,
$$P + Q = 1. \tag{24}$$
Thus, the abstract state $\boldsymbol{s}(D)$ of $\boldsymbol{u}(D)$ can be written concisely as
$$\boldsymbol{s}(D) = \boldsymbol{u}(D)PG(D)Q. \tag{25}$$
Note that in an observer canonical form of a polynomial encoding matrix the abstract states are in 1-1 correspondence with the physical states since the contents of the memory elements are simply shifted out in the absence of inputs.

Since the abstract state does not depend on the realization we have the same abstract states in the observer canonical form as in the controller canonical form.

IV. Equivalence of Encoders and Basic Encoders

In a communication context it is natural to say that two encoders are equivalent if they generate the same code C. It is therefore important to look for encoders with the lowest complexity within the class of equivalent encoders.

Definition Two convolutional encoding matrices $G(D)$ and $G'(D)$ are called *equivalent* if they encode the same code. Two convolutional encoders are called *equivalent* if their encoding matrices are equivalent.

Theorem 3 *Two rate $R = b/c$ convolutional encoding matrices $G(D)$ and $G'(D)$ are equivalent, if and only if there is a $b \times b$ nonsingular matrix $T(D)$ over $\boldsymbol{F}_2(D)$ such that*

$$G(D) = T(D)G'(D). \tag{26}$$

Proof If (26) holds, then clearly $G(D)$ and $G'(D)$ are equivalent.

Conversely, suppose that $G(D)$ and $G'(D)$ are equivalent. Let $\boldsymbol{g}_i(D) \in \boldsymbol{F}_2(D)^{(c)}$ be the ith row of $G(D)$. Then there exists a $\boldsymbol{u}_i(D) \in \boldsymbol{F}_2((D))^{(b)}$ such that

$$\boldsymbol{g}_i(D) = \boldsymbol{u}_i(D)G'(D). \tag{27}$$

Let

$$T(D) = \begin{pmatrix} \boldsymbol{u}_1(D) \\ \boldsymbol{u}_2(D) \\ \vdots \\ \boldsymbol{u}_b(D) \end{pmatrix}. \tag{28}$$

Then

$$G(D) = T(D)G'(D), \tag{29}$$

where $T(D)$ is a $b \times b$ matrix over $\boldsymbol{F}_2((D))$. Let $S'(D)$ be a $b \times b$ nonsingular submatrix of $G'(D)$ and $S(D)$ be the corresponding $b \times b$ submatrix of $G(D)$. Then $S(D) = T(D)S'(D)$. Thus $T(D) = S(D)S'(D)^{-1}$ and, hence, $T(D)$ is over $\boldsymbol{F}_2(D)$. Since $G(D)$, being an encoding matrix, is of rank b it follows that $T(D)$ is also of rank b and, hence, is nonsingular. □

Example 2 By Theorem 3 the encoding matrix for the rate $R = 2/3$ convolutional encoder shown in Fig. 6

$$G(D) = \begin{pmatrix} 1+D & D & 1 \\ D^2 & 1 & 1+D+D^2 \end{pmatrix}$$

is equivalent to the encoding matrix

$$G'(D) = \begin{pmatrix} 1+D & D & 1 \\ 1+D^2+D^3 & 1+D+D^2+D^3 & 0 \end{pmatrix},$$

since there is a nonsingular matrix

$$T(D) = \begin{pmatrix} 1 & 0 \\ 1+D+D^2 & 1 \end{pmatrix}$$

such that $G'(D) = T(D)G(D)$. The controller canonical form of $G'(D)$ is shown in Fig. 7. Since $G(D)$ and $G'(D)$ are equivalent, the encoders in Figs. 6 and 7 encode the same code.

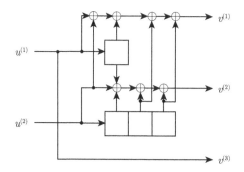

Fig. 7 Controller canonical form of the encoding matrix $G'(D)$

Definition (Forney [1]) A convolutional encoding matrix is called *basic* if it is polynomial and it has a polynomial right inverse. A convolutional encoder is called *basic* if its encoding matrix is basic.

Next, we consider the invariant-factor decomposition of a rational matrix [1], [5].

Invariant-Factor Theorem Let $G(D)$ be a $b \times c, b \leqslant c$, binary rational matrix of rank r. Then $G(D)$ can be written in the following manner:

$$G(D) = A(D)\Gamma(D)B(D), \tag{30}$$

where $A(D)$ and $B(D)$ are $b \times b$ and $c \times c$, respectively, binary unimodular matrices and where $\Gamma(D)$ is the $b \times c$ matrix

$$\Gamma(D) = \begin{pmatrix} \frac{\alpha_1(D)}{\beta_1(D)} & & & & & \\ & \frac{\alpha_2(D)}{\beta_2(D)} & & & & \\ & & \ddots & & & \\ & & & \frac{\alpha_r(D)}{\beta_r(D)} & & 0^{(b,c-b)} \\ & & & & 0 & \\ & & & & & \ddots \\ & & & & & & 0 \end{pmatrix},$$

in which $0^{(b,c-b)}$ is a $b \times (c-b)$ zero matrix, $\alpha_i(D)$ and $\beta_i(D)$ are nonzero binary polynomials satisfying $\alpha_i(D)|\alpha_{i+1}(D)$ and $\beta_{i+1}(D)|\beta_i(D), i = 1, 2, \cdots, r-1$.

A square binary polynomial matrix is called *unimodular* if its determinant is 1. The unimodular matrices are uniquely characterized as the square polynomial matrices that have polynomial inverses. In an invariant-factor decomposition the unimodular matrices $A(D)$ and $B(D)$ are not unique, although the $\frac{\alpha_i(D)}{\beta_i(D)}$'s are uniquely determined by $G(D)$ and are called the invariant factors of $G(D)$.

Consider a rational encoding matrix $G(D)$ with invariant-factor decomposition $G(D) = A(D)\Gamma(D)B(D)$ and let $G'(D)$ be an encoding matrix consisting of the first b rows of $B(D)$. Then

$$G(D) = A(D) \begin{pmatrix} \frac{\alpha_1(D)}{\beta_1(D)} & & & \\ & \frac{\alpha_2(D)}{\beta_2(D)} & & \\ & & \ddots & \\ & & & \frac{\alpha_b(D)}{\beta_b(D)} \end{pmatrix} G'(D), \quad (31)$$

where $\alpha_i(D)$ and $\beta_i(D)$ are polynomials satisfying $\alpha_i(D)|\alpha_{i+1}(D)$ and $\beta_{i+1}(D)|$

$\beta_i(D)$, $i = 1, 2, \cdots, b-1$. Since both $A(D)$ and

$$\begin{pmatrix} \frac{\alpha_1(D)}{\beta_1(D)} & & & \\ & \frac{\alpha_2(D)}{\beta_2(D)} & & \\ & & \ddots & \\ & & & \frac{\alpha_b(D)}{\beta_b(D)} \end{pmatrix}$$

are nonsingular matrices over $\boldsymbol{F}_2(D)$ it follows from Theorem 3 that $G(D)$ and $G'(D)$ are equivalent. But $G'(D)$ is polynomial and since $B(D)$ has a polynomial inverse, it follows that $G'(D)$ has a polynomial right inverse (consisting of the first b columns of $B^{-1}(D)$). Therefore, $G'(D)$ is basic and we have the following algorithm.

An Algorithm to Construct a Basic Encoding Matrix Equivalent to a Given Encoding Matrix:

1) Compute the invariant-factor decomposition $G(D) = A(D)\Gamma(D)B(D)$.

2) Let $G'(D)$ be the first b rows of $B(D)$. $G'(D)$ is a basic encoding matrix equivalent to $G(D)$.

We summarize these results in Theorem 4.

Theorem 4 (Forney [1])   *Every rational encoding matrix is equivalent to a basic convolutional encoding matrix.*

Now, we have the following.

Theorem 5 (Forney [1])   *Tow basic convolutional encoding matrices $G(D)$ and $G'(D)$ are equivalent if and only if $G'(D) = T(D)G(D)$, where $T(D)$ is a $b \times b$ polynomial matrix with determinant 1.*

Proof   Let $G'(D) = T(D)G(D)$, where $T(D)$ is a polynomial matrix with determinant 1. By Theorem 3, $G(D)$ and $G'(D)$ are equivalent.

Conversely, suppose that $G'(D)$ and $G(D)$ are equivalent. By Theorem 3, there is a nonsingular $b \times b$ matrix $T(D)$ over $\boldsymbol{F}_2(D)$ such that $G'(D) = T(D)G(D)$. Since $G(D)$ is basic it has a polynomial right inverse $G^{-1}(D)$. Then, $T(D) = G'(D)G^{-1}(D)$ is polynomial. We can repeat the argument with

$G(D)$ and $G'(D)$ reversed to obtain $G(D) = S(D)G'(D)$ for some polynomial matrix $S(D)$. Thus, $G(D) = S(D)T(D)G(D)$. Since $G(D)$ is of full rank, we conclude that $S(D)T(D) = I_b$. Finally, since both $T(D)$ and $S(D)$ are polynomial, $T(D)$ must have determinant 1 and the proof is completed. □

Let $G(D) = A(D)\Gamma(D)B(D)$ be an invariant-factor decomposition of a basic encoding matrix $G(D)$. Then $G(D) = A(D)G'(D)$ where $G'(D)$ is the $b \times c$ polynomial matrix that consists of the first b rows of the matrix $B(D)$. Since the matrix $A(D)$ is a $b \times b$ unimodular matrix it follows from Theorem 5 that $G(D)$ and $G'(D)$ are equivalent basic encoding matrices.

V. Minimal-Basic Encoding Matrices

We shall now show that among all equivalent encoding matrices there exists a basic encoding matrix whose controller canonical form requires a minimal number of memory elements.

First, we shall consider only basic encoding matrices. The following definition is equivalent to Forney's definition of a minimal encoder [1].

Definition A *minimal-basic* encoding matrix is a basic encoding matrix whose overall constraint length ν is minimal over all equivalent basic encoding matrices.

Let $G(D)$ be a basic encoding matrix. The positions for the row-wise highest order coefficients in $G(D)$ will play a significant role in the sequel. Hence, we let $[G(D)]_h$ be a $(0,1)$-matrix with 1 in the position (i,j) where $\deg g_{ij}(D) = \nu_i$ and 0, otherwise.

Let us write
$$G(D) = G_0(D) + G_1(D), \qquad (32)$$

where

$$G_1(D) = \begin{pmatrix} D^{\nu_1} & & & \\ & D^{\nu_2} & & \\ & & \ddots & \\ & & & D^{\nu_b} \end{pmatrix} [G(D)]_h. \qquad (33)$$

Then, all entries in the ith row of $G_0(D)$ are of degree $< \nu_i$. Clearly the maximum degree μ among the $b \times b$ subdeterminants of $G(D)$ is $\leqslant \nu$.

Theorem 6 Let $G(D)$ be a $b \times c$ basic encoding matrix with overall constraint length ν. Then, the following statements are equivalent.

a) $G(D)$ is a minimal-basic encoding matrix.

b) The maximum degree μ among the $b \times b$ subdeterminants of $G(D)$ is equal to the overall constraint length ν.

c) $[G(D)]_h$ is of full rank.

Proof It follows immediately from (33) that b) and c) are equivalent. Thus we need only prove that a) and b) are equivalent.

a)$\Rightarrow$b): Assume that $G(D)$ is minimal-basic.

Suppose that $\mu < \nu$, i.e., rank $[G(D)]_h < b$. Denote the rows of $G(D)$ by $r_1, r_2, \cdots, r_b$ and the rows of $[G(D)]_h$ by $[r_1], [r_2], \cdots, [r_b]$. Then there is a linear relation

$$[r_{i_1}] + [r_{i_2}] + \cdots + [r_{i_d}] = \mathbf{0}. \tag{34}$$

The ith row of $G_1(D)$ is $D^{\nu_i}[r_i]$. Without loss of generality we can assume that $\nu_{i_d} \geqslant \nu_{i_j}, j = 1, 2, \cdots, d-1$. Adding

$$D^{\nu_{i_d}-\nu_{i_1}} D^{\nu_{i_1}}[r_{i_1}] + D^{\nu_{i_d}-\nu_{i_2}} D^{\nu_{i_2}}[r_{i_2}] + \cdots$$
$$+ D^{\nu_{i_d}-\nu_{i_{d-1}}} D^{\nu_{i_{d-1}}}[r_{i_{d-1}}]$$
$$= D^{\nu_{i_d}}([r_{i_1}] + [r_{i_2}] + \cdots + [r_{i_{d-1}}]) \tag{35}$$

to the i_dth row of $G_1(D)$ reduces it to an all zero row. Similarly, adding

$$D^{\nu_{i_d}-\nu_{i_1}} r_1 + D^{\nu_{i_d}-\nu_{i_2}} r_2 + \cdots + D^{\nu_{i_d}-\nu_{i_{d-1}}} r_{i_{d-1}} \tag{36}$$

to the i_dth row of $G(D)$ will reduce the highest degree of the i_dth row of $G(D)$ but leave the other rows of $G(D)$ unchanged. Thus we obtain a basic encoding matrix equivalent to $G(D)$ with an overall constraint length which is less than that of $G(D)$. This is a contradiction to the assumption that $G(D)$ is minimal-basic and we conclude that $\mu = \nu$.

b)⇒ a): Assume that $\mu = \nu$.

Let $G'(D)$ be a basic encoding matrix equivalent to $G(D)$. From Theorem 5 follows that $G'(D) = T(D)G(D)$, where $T(D)$ is a $b \times b$ polynomial matrix with determinant 1. Since $\det T(D) = 1$, the maximum degree among the $b \times b$ subdeterminants of $G'(D)$ is equal to that of $G(D)$. Hence, μ is invariant over all equivalent basic encoding matrices.

Clearly μ is less than or equal to the overall constraint length for all equivalent basic encoding matrices and it follows that $G(D)$ is a minimal-basic encoding matrix. □

Corollary 7 *Let $G(D)$ be a $b \times c$ basic encoding matrix with maximum degree μ among its $b \times b$ subdeterminants. Then $G(D)$ has an equivalent minimal-basic encoding matrix whose overall constraint length $\nu = \mu$.*

Proof Follows from the proof of Theorem 6 and the fact that μ is invariant over all equivalent basic encoding matrices. □

Example 3 Consider the encoding matrix for the encoder in Fig. 7, viz.,

$$G'(D) = \begin{pmatrix} 1+D & D & 1 \\ 1+D^2+D^3 & 1+D+D^2+D^3 & 0 \end{pmatrix}.$$

The rank of

$$[G'(D)]_h = \begin{pmatrix} 1 & 1 & 0 \\ 1 & 1 & 0 \end{pmatrix}$$

is one. Hence, $G'(D)$ cannot be a minimal-basic encoding matrix.

On the other hand, $G'(D)$ has the following three $b \times b$ subdeterminants:

$$1+D+D^3, 1+D^2+D^3, 1+D+D^2+D^3$$

and, thus, $\mu = 3$. Hence, any minimal-basic encoding matrix equivalent to $G'(D)$ has overall constraint length $\nu = 3$.

The equivalent basic encoding matrix for the encoder in Fig. 6

$$G(D) = \begin{pmatrix} 1+D & D & 1 \\ D^2 & 1 & 1+D+D^2 \end{pmatrix}$$

has
$$[G(D)]_h = \begin{pmatrix} 1 & 1 & 0 \\ 1 & 0 & 1 \end{pmatrix}$$

with full rank and, hence, is a minimal-basic encoding matrix.

Clearly we can use the technique in the proof of Theorem 6 to obtain a minimal-basic encoding matrix equivalent to the basic encoding matrix $G'(D)$ for the encoder in Fig. 7. We simply multiply the first row of $G'(D)$ by $D^{\nu_2-\nu_1} = D^2$ and add it to the second row:

$$\begin{pmatrix} 1 & 0 \\ D^2 & 1 \end{pmatrix} \begin{pmatrix} 1+D & D & 1 \\ 1+D^2+D^3 & 1+D+D^2+D^3 & 0 \end{pmatrix}$$
$$= \begin{pmatrix} 1+D & D & 1 \\ 1 & 1+D+D^2 & D^2 \end{pmatrix}.$$

Thus, the minimal-basic encoding matrix equivalent to a given basic encoding matrix is not necessarily unique.

In general, we have [1] the next algorithm.

A Simple Algorithm to Construct a Minimal-Basic Encoding Matrix Equivalent to a Given Basic Encoding Matrix:

1) IF $[G(D)]_h$ has full rank, THEN $G(D)$ is a minimal-basic encoding matrix and we STOP; ELSE GO TO next step.

2) Let $[r_{i_1}], [r_{i_2}], \cdots, [r_{i_d}]$ denote a set of rows of $[G(D)]_h$ such that $\nu_{i_d} \geqslant \nu_{i_j}, 1 \leqslant j < d$, and
$$[r_{i_1}] + [r_{i_2}] + \cdots + [r_{i_d}] = \mathbf{0}.$$

Let $r_{i_1}, r_{i_2}, \cdots, r_{i_d}$ denote the corresponding set of rows of $G(D)$. Add
$$D^{\nu_{i_d}-\nu_{i_1}}r_1 + D^{\nu_{i_d}-\nu_{i_2}}r_2 + \cdots + D^{\nu_{i_d}-\nu_{i_{d-1}}}r_{i_{d-1}}$$

to the i_dth row of $G(D)$.

Call the new matrix $G(D)$ and GO TO step 1.

Corollary 8 (Forney [1])   *Every encoding matrix is equivalent to a minimal-basic encoding matrix.*

The physical state space of a controller canonical form of an encoding matrix of overall constraint length ν contains 2^ν states. This type of realization plays an important role in connection with minimal-basic encoding matrices as we shall see in the sequel, but first we prove a technical lemma.

Lemma 9 *Let $G(D)$ be a minimal-basic encoding matrix and let*

$$\boldsymbol{u}(D) = \sum_{i=-m}^{n} (u_i^{(1)} u_i^{(2)} \cdots u_i^{(b)}) D^i, \tag{37}$$

where m is the memory of $G(D)$ and $n \geq -m$. If $\boldsymbol{u}(D)G(D)Q = 0$, then $\boldsymbol{u}(D) = \boldsymbol{0}$.

Proof Let

$$\boldsymbol{v}(D) = \boldsymbol{u}(D)G(D). \tag{38}$$

Then, by assumption,

$$\boldsymbol{v}(D)Q = \boldsymbol{u}(D)G(D)Q = \boldsymbol{0}. \tag{39}$$

Thus each coefficient of $D^i, i \geq 0$, in $\boldsymbol{v}(D)$ must be 0. Write $G(D)$ as in (32), then we have

$$\begin{aligned}\boldsymbol{v}(D) =& \boldsymbol{u}(D)G_0(D) \\ &+ \boldsymbol{u}(D)\begin{pmatrix} D^{\nu_1} & & & \\ & D^{\nu_2} & & \\ & & \ddots & \\ & & & D^{\nu_b} \end{pmatrix}[G(D)]_h. \end{aligned} \tag{40}$$

Without loss of generality, we can assume that

$$m = \nu_1 = \nu_2 = \cdots = \nu_l > \nu_{l=1} \geq \cdots \geq \nu_b. \tag{41}$$

Then the coefficient of D^{m+n}, where $m+n \geq 0$, in $\boldsymbol{v}(D)$ is

$$(u_n^{(1)} u_n^{(2)} \cdots u_n^{(l)} 0 \cdots 0)[G(D)]_h,$$

which must be $\boldsymbol{0}$. Since $G(D)$ is a minimal-basic encoding matrix, $[G(D)]_h$ is of rank b. Hence, $u_n^{(1)} = u_n^{(2)} = \cdots = u_n^{(l)} = 0$. Proceeding in this way, we can prove that $\boldsymbol{u}(D) = \boldsymbol{0}$. $\square$

(Lemma 9 also follows form the *predictable degree property* introduced in [1].)

Theorem 10 Let $G(D)$ be a minimal-basic encoding matrix whose overall constraint length is ν. Then[2]

$$\#\{\text{abstract states}\} = 2^\nu. \tag{42}$$

Proof Consider the controller canonical form of the minimal-basic encoding matrix $G(D)$. Clearly input sequences of the form

$$\boldsymbol{u}(D) = \left(\sum_{i=1}^{\nu_1} u_{-i}^{(1)} D^{-i} \sum_{i=1}^{\nu_2} u_{-i}^{(2)} D^{-i} \cdots \sum_{i=1}^{\nu_b} u_{-i}^{(b)} D^{-i} \right) \tag{43}$$

will carry us to all physical states at time 0. Then we have the abstract states

$$\boldsymbol{s}(D) = \boldsymbol{u}(D) G(D) Q, \tag{44}$$

where $\boldsymbol{u}(D)$ is of the form given in (43). Every abstract state can be obtained in this way and we have

$$\#\{\text{abstract states}\} \leqslant 2^\nu. \tag{45}$$

To prove that the equality sign holds, it is enough to show that $\boldsymbol{u}(D) = \boldsymbol{0}$ is the only physical state that produces the abstract state $\boldsymbol{s}(D) = \boldsymbol{0}$. This follows from Lemma 9. □

We shall conclude this section by proving that the constraint lengths are invariants of equivalent minimal-basic encoding matrices. First we need the following

Lemma 11 Let V be a k-dimensional vector space over a field F and let $\{\boldsymbol{\alpha}_1, \boldsymbol{\alpha}_2, \cdots, \boldsymbol{\alpha}_k\}$ be a basis of V. Let $\{\boldsymbol{\beta}_1, \boldsymbol{\beta}_2, \cdots, \boldsymbol{\beta}_l\}$ be a set of l, $l < k$, linearly independent vectors of V. Then there exist $k - l$ vectors $\boldsymbol{\alpha}_{i_{l+1}}, \boldsymbol{\alpha}_{i_{l+2}}, \cdots, \boldsymbol{\alpha}_{i_k}$, $1 \leqslant i_{l+1} < i_{l+2} < \cdots < i_k \leqslant k$, such that $\{\boldsymbol{\beta}_1, \boldsymbol{\beta}_2, \cdots, \boldsymbol{\beta}_l, \boldsymbol{\alpha}_{i_{l+1}}, \boldsymbol{\alpha}_{i_{l+2}}, \cdots, \boldsymbol{\alpha}_{i_k}\}$ is also a basis of V.

---

2 $\#\{\cdot\}$ denotes the cardinality of the set $\{\cdot\}$.

Proof Consider the vectors in the sequence $\boldsymbol{\beta}_1, \boldsymbol{\beta}_2, \cdots, \boldsymbol{\beta}_l, \boldsymbol{\alpha}_1, \boldsymbol{\alpha}_2, \cdots, \boldsymbol{\alpha}_k$ one by one successively from left to right. If the vector under consideration is a linear combination of vectors to the left of it, then delete it; otherwise keep it. Finally, we obtain a basis $\boldsymbol{\beta}_1, \boldsymbol{\beta}_2, \cdots, \boldsymbol{\beta}_l, \boldsymbol{\alpha}_{i_{l+1}}, \boldsymbol{\alpha}_{i_{l+2}}, \cdots, \boldsymbol{\alpha}_{i_k}, 1 \leqslant i_{l+1} < \cdots < i_k \leqslant k$, of V. □

Theorem 12 *The constraint lengths of two equivalent minimal-basic encoding matrices are equal one by one up to a rearrangement.*

Proof Let $G(D)$ and $G'(D)$ be two equivalent minimal-basic encoding matrices with constraint lengths $\nu_1, \nu_2, \cdots, \nu_b$ and $\nu'_1, \nu'_2, \cdots, \nu'_b$, respectively. Without loss of generality, we assume that $\nu_1 \leqslant \nu_2 \leqslant \cdots \leqslant \nu_b$ and $\nu'_1 \leqslant \nu'_2 \leqslant \cdots \leqslant \nu'_b$.

Now suppose that ν_i and ν'_i are not equal for all $i, 1 \leqslant i \leqslant b$. Let j be the smallest index such that $\nu_j \neq \nu'_j$. Then, without loss of generality, we assume that $\nu_j < \nu'_j$. From the sequence $\boldsymbol{g}_1, \boldsymbol{g}_2, \cdots, \boldsymbol{g}_j, \boldsymbol{g}'_1, \boldsymbol{g}'_2, \cdots, \boldsymbol{g}'_b$ we can according to Lemma 11, obtain a basis $\boldsymbol{g}_1, \boldsymbol{g}_2, \cdots, \boldsymbol{g}_j, \boldsymbol{g}'_{i_{j+1}}, \boldsymbol{g}'_{i_{j+2}}, \cdots, \boldsymbol{g}'_{i_b}$ of C. These b row vectors form an encoding matrix $G''(D)$ which is equivalent to $G'(D)$. Let

$$\{g'_1, g'_2, \cdots, g'_b\} \setminus \{g'_{i_{j+1}}, g'_{i_{j+2}}, \cdots, g'_{i_b}\} = \{g'_{i_1}, g'_{i_2}, \cdots, q'_{i_j}\}. \tag{46}$$

From our assumptions, it follow that

$$\sum_{l=1}^{j} \nu_i < \sum_{l=1}^{j} \nu'_i \leqslant \sum_{l=1}^{j} \nu'_{i_l}. \tag{47}$$

Then we have

$$\nu'' = \sum_{l=1}^{j} \nu_i + \sum_{l=j+1}^{b} \nu'_{i_l} < \sum_{l=1}^{j} \nu'_{i_l} + \sum_{l=j+1}^{b} \nu'_{i_l} = \nu', \tag{48}$$

where ν' and ν'' are the overall constraint lengths of the encoding matrices $G'(D)$ and $G''(D)$, respectively.

From Theorem 3, it follows that there exists a $b \times b$ nonsingular matrix $T(D)$ over $\boldsymbol{F}_2(D)$ such that

$$G''(D) = T(D) G'(D). \tag{49}$$

Since $G''(D)$ is basic it has a polynomial right inverse $G'^{-1}(D)$ and it follows that
$$T(D) = G''(D)G'^{-1}(D) \qquad (50)$$
is polynomial. Denote by μ' and μ'' the maximum degrees among the $b \times b$ minors of $G'(D)$ and $G''(D)$, respectively. It follows from (49) that
$$\mu'' = \deg |T(D)| + \mu'. \qquad (51)$$
Clearly $\nu'' \geqslant \mu''$ and, since $G'(D)$ is minimal-basic, $\nu' = \mu'$ by Theorem 6. Thus,
$$\nu'' \geqslant \deg|T(D)| + \nu' \geqslant \nu', \qquad (52)$$
which contradicts (48) and the proof is completed. □

Corollary 13 *Two equivalent minimal-basic encoding matrices have the same memory.*

The statement in Theorem 12 is equivalent to a classical result of Kronecker [6]; see also Forney [7].

VI. Minimal Encoding Matrices

We shall now proceed to show that a minimal-basic encoding matrix is also minimal in a more general sense.

Definition A convolutional encoding matrix is *minimal* if its number of abstract states is minimal over all equivalent encoding matrices.

Before we can show that every minimal-basic encoding matrix is also a (basic) minimal encoding matrix we have to prove the following lemmas.

Lemma 14 (Forney [1],[3]) *Only the zero abstract state of a minimal-basic encoding matrix $G(D)$ can be a codeword.*

Proof We can assume that the abstract state $\boldsymbol{s}(D)$ arises from an input $\boldsymbol{u}(D)$ which is polynomial in D^{-1} and of degree $\leqslant m$ and without a constant term, i.e., $\boldsymbol{u}(0) = 0$. Thus,
$$\boldsymbol{s}(D) = \boldsymbol{u}(D)G(D)Q. \qquad (53)$$

Then, it follows that

$$u(D)G(D) = w(D) + s(D), \tag{54}$$

where $w(D)$ is polynomial in D^{-1} without a constant term.

Assume that $s(D)$ is a codeword, i.e., there is an input $u'(D) \in F_2((D))$ such that

$$s(D) = u'(D)G(D). \tag{55}$$

Since $s(D)$ is polynomial and $G(D)$ has a polynomial inverse it follows that $u'(D) \in F_2[D]$.

Combining (54) and (55) we have

$$(u(D) + u'(D))G(D) = w(D). \tag{56}$$

Consequently

$$(u(D) + u'(D))G(D)Q = 0. \tag{57}$$

By Lemma 9,

$$u(D) + u'(D) = 0 \tag{58}$$

and, hence, $u'(D) = 0$. It follows from (55) that $s(D) = 0$. □

Lemma 15 *Let $G(D)$ and $G'(D)$ be equivalent encoding matrices. Then, every abstract state of $G(D)$ can be expressed as a sum of an abstract state of $G'(D)$ and a codeword. Furthermore, if $G'(D)$ is minimal-basic, then the expression is unique.*

Proof Assume that $G(D) = T(D)G'(D)$, where $T(D)$ is a $b \times b$ nonsingular matrix over $F_2(D)$. Any abstract state of $G(D)$, $s_G(D)$, can be written in the form $u(D)G(D)Q$, where $u(D)$ is polynomial in D^{-1} without a constant term. Thus, we have

$$\begin{aligned}s_G(D) &= u(D)G(D)Q = u(D)T(D)G'(D)Q \\ &= u(D)T(D)(P+Q)G'(D)Q \\ &= u(D)T(D)PG'(D)Q + u(D)T(D)QG'(D)Q.\end{aligned} \tag{59}$$

Since $\boldsymbol{u}(D)T(D)P$ is polynomial in D^{-1} without a constant term it follows from (25) that
$$\boldsymbol{s}_{G'}(D) = \boldsymbol{u}(D)T(D)PG'(D)Q \tag{60}$$
is an abstract state of $G(D)$. Furthermore, $\boldsymbol{u}(D)T(D)Q$ is a formal power series, and so is $\boldsymbol{u}(D)T(D)QG'(D)$. Hence,
$$\boldsymbol{v}(D) \stackrel{\text{def}}{=} \boldsymbol{u}(D)T(D)QG'(D)Q = \boldsymbol{u}(D)T(D)QG'(D) \tag{61}$$
is a codeword encoded by $G'(D)$. Combining (59), (60), and (61) we obtain
$$\boldsymbol{s}_G(D) = \boldsymbol{s}_{G'}(D) + \boldsymbol{v}(D) \tag{62}$$
and we have proved that every abstract state of $G(D)$ can be written as a sum of an abstract state of $G'(D)$ and a codeword.

Assume now that $G'(D) = G_{mb}(D)$ is minimal-basic. To prove uniqueness, we assume that
$$\boldsymbol{s}_G(D) = \boldsymbol{s}_{mb}(D) + \boldsymbol{v}(D) = \boldsymbol{s}'_{mb}(D) + \boldsymbol{v}'(D), \tag{63}$$
where $\boldsymbol{s}_{mb}(D), \boldsymbol{s}'_{mb}(D)$ are abstract states of $G_{mb}(D)$, and $\boldsymbol{v}(D), \boldsymbol{v}'(D)$ are codewords. Since the sum of two abstract states is an abstract state and the sum of two codewords is a codeword it follows from (63) that
$$\boldsymbol{s}''_{mb}(D) = \boldsymbol{s}_{mb}(D) + \boldsymbol{s}'_{mb}(D) = \boldsymbol{v}(D) + \boldsymbol{v}'(D) = \boldsymbol{v}''(D) \tag{64}$$
is both an abstract state of $G_{mb}(D)$ and a codeword. From Lemma 11 we deduce that
$$\boldsymbol{s}''_{mb}(D) = \boldsymbol{0} \tag{65}$$
and, hence, that
$$\boldsymbol{s}_{mb}(D) = \boldsymbol{s}'_{mb}(D), \tag{66}$$
and
$$\boldsymbol{v}(D) = \boldsymbol{v}'(D), \tag{67}$$

which completes the proof. □

Theorem 16 (Forney [1]) Let $G(D)$ be any encoding matrix equivalent to a minimal-basic encoding matrix $G_{mb}(D)$. Then

$$\#\{\text{abstract states of } G(D)\}$$
$$\geq \#\{\text{abstract states of } G_{mb}(D)\}. \quad (68)$$

Proof Consider the following map:

$$\phi : \{\text{abstract states of } G(D)\} \to \{\text{abstract states of } G_{mb}(D)\}$$
$$s_G(D) \mapsto s_{mb}(D),$$

where

$$s_G(D) = s_{mb}(D) + v(D), \quad (69)$$

in which $v(D)$ is a codeword. From Lemma 15, it follows that ϕ is well-defined.

By the first statement of Lemma 15 we can prove that every abstract state $s_{mb}(D)$ can be written as a sum of an abstract state of $G(D)$ and a codeword. Hence, we conclude that ϕ is surjective which proves the theorem. □

Remark The map ϕ in Theorem 16 is clearly linear. Moreover, if $G(D)$ is a minimal encoding matrix, then ϕ is necessarily an isomorphism of abstract state space of $G(D)$ and that of $G_{mb}(D)$.

From Theorem 16, Corollary 17 follows immediately.

Corollary 17 Every minimal-basic encoding matrix is a (basic) minimal encoding matrix.

Next, we shall prove the following little lemma.

Lemma 18 Let $G(D)$ be a $b \times c$ matrix of rank b whose entries are rational functions of D. Then a necessary and sufficient condition for $G(D)$ to have a polynomial inverse is: for each $u(D) \in F_2^b(D)$ satisfying $u(D)G(D) \in F_2^c[D]$ we must have $u(D) \in F_2^b[D]$.

Proof Since the necessity of the condition is obvious we shall prove only the sufficiency. Let us assume that $G(D)$ does not have a polynomial inverse.

Then, from the invariant-factor decomposition

$$G(D) = A(D) \begin{pmatrix} \frac{\alpha_1(D)}{\beta_1(D)} & & & & & \\ & \frac{\alpha_2(D)}{\beta_2(D)} & & & & \\ & & \ddots & & & \\ & & & \frac{\alpha_b(D)}{\beta_b(D)} & 0 & \cdots & 0 \end{pmatrix} B(D) \quad (70)$$

follows that $\alpha_b(D) \neq 1$. Clearly,

$$\boldsymbol{u}(D) = \left(0 \cdots 0 \frac{\beta_b(D)}{\alpha_b(D)} \right) A^{-1}(D) \notin \boldsymbol{F}_2^b[D] \quad (71)$$

but

$$\begin{aligned} \boldsymbol{u}(D)G(D) &= \left(0 \cdots 0 \frac{\beta_b(D)}{\alpha_b(D)} \right) A^{-1}(D) G(D) \\ &= (0 \cdots 0 1 0 \cdots 0) B(D) \in \boldsymbol{F}_2^c[D]. \end{aligned} \quad (72)$$

Hence, we have proved our lemma. □

We are now well prepared to prove the following theorem on minimal encoding matrices.

Theorem 19 (cf. [1]) *Let $G(D)$ be an encoding matrix and $G_{mb}(D)$ be an equivalent minimal-basic encoding matrix. Then, the following statements are equivalent.*

a) $G(D)$ is a minimal encoding matrix.

b) #{abstract states of $G(D)$} = #{abstract states of $G_{mb}(D)$}.

c) Only the zero abstract state of $G(D)$ can be a codeword.

d) $G(D)$ has a polynomial right inverse in D and a polynomial right inverse in D^{-1}.

Proof It follows immediately from Theorem 16 that a) and b) are equivalent.

Next, we prove that b) and c) are equivalent. In the proof of Theorem 16, we have defined a surjective map

$$\phi : \{\text{abstract states of } G(D)\} \to \{\text{abstract states of } G_{mb}(D)\}$$

$$s_G(D) \mapsto s_{mb}(D),$$

where
$$s_G(D) = s_{mb}(D) + v(D), \qquad (73)$$

in which $v(D)$ is a codeword. Clearly, ϕ is injective, if and only if b) holds, and if and only if c) holds. Hence, b) and c) are equivalent.

It remains to prove that c) and d) are equivalent.

c)$\Rightarrow$d): Suppose that c) holds. First we shall prove that $G(D)$ has a polynomial right inverse in D^{-1}. Let $\boldsymbol{u}(D) \in \boldsymbol{F}_2^b(D)$ and assume that $\boldsymbol{v}(D) = \boldsymbol{u}(D)G(D)$ is polynomial in D^{-1}. Then $D^{-1}\boldsymbol{v}(D)$ is polynomial in D^{-1} without a constant term, i.e.,

$$D^{-1}\boldsymbol{v}(D)Q = \boldsymbol{0}. \qquad (74)$$

But
$$\begin{aligned} D^{-1}\boldsymbol{v}(D)Q &= D^{-1}\boldsymbol{u}(D)(P+Q)G(D)Q \\ &= D^{-1}\boldsymbol{u}(D)PG(D)Q + D^{-1}\boldsymbol{u}(D)QG(D)Q = \boldsymbol{0}, \end{aligned} \qquad (75)$$

where
$$D^{-1}\boldsymbol{u}(D)QG(D)Q = D^{-1}\boldsymbol{u}(D)QG(D) \qquad (76)$$

is a codeword. Hence, from (75) and (76) it follows that the abstract state $D^{-1}\boldsymbol{u}(D)PG(D)Q$ is a codeword and, then, since c) holds, it is the zero codeword. Thus,

$$D^{-1}\boldsymbol{u}(D)QG(D) = \boldsymbol{0} \qquad (77)$$

and, since $G(D)$ has full rank,

$$D^{-1}\boldsymbol{u}(D)Q = \boldsymbol{0} \qquad (78)$$

or, in other words, $\boldsymbol{u}(D)$ is polynomial in D^{-1}. Since every rational function in D can be written as a rational function in D^{-1}, $G(D)$ can be written as a matrix whose entries are rational functions in D^{-1}. We can apply Lemma 18 and conclude that $G(D)$ has a polynomial right inverse in D^{-1}.

Next, we shall prove that $G(D)$ has a polynomial right pseudo-inverse in D. (By a polynomial right pseudo-inverse of $G(D)$ we mean a $c \times b$ polynomial matrix $\widetilde{G^{-1}}(D)$ such that $G(D)\widetilde{G^{-1}}(D) = D^s I_b$ for some $s \geqslant 0$.) Let $G_{-1}^{-1}(D)$ be a polynomial right inverse in D^{-1} of $G(D)$. Then there exists an integer $s \geqslant 0$ such that $D^s G_{-1}^{-1}(D)$ is a polynomial matrix in D and

$$G(D)D^s G_{-1}^{-1}(D) = D^s I_b, \tag{79}$$

i.e., $D^s G_{-1}^{-1}(D)$ is a polynomial right pseudo-inverse in D of $G(D)$.

Finally, we shall prove that $G(D)$ has also a polynomial right inverse in D. Let $\boldsymbol{u}(D) \in \boldsymbol{F}_2^b(D)$ and assume that $\boldsymbol{v}(D) = \boldsymbol{u}(D)G(D)$ is polynomial in D. Then,

$$\boldsymbol{v}(D) = \boldsymbol{u}(D)PG(D) + \boldsymbol{u}(D)QG(D), \tag{80}$$

where $\boldsymbol{u}(D)Q$ is a formal power series. Thus, $\boldsymbol{u}(D)QG(D)$ is also a formal power series and, since $\boldsymbol{v}(D)$ is polynomial in D, it follows that $\boldsymbol{u}(D)PG(D)$ is a formal power series. Then

$$\boldsymbol{u}(D)PG(D) = \boldsymbol{u}(D)PG(D)Q \tag{81}$$

is an abstract state. From (80), it follows that it is also a codeword and, since c) holds, we conclude that

$$\boldsymbol{u}(D)PG(D) = \boldsymbol{0}. \tag{82}$$

Since $G(D)$ has full rank,

$$\boldsymbol{u}(D)P = \boldsymbol{0} \tag{83}$$

or, in other words, $\boldsymbol{u}(D)$ is a formal power series.

Since $\boldsymbol{v}(D)$ is polynomial and $D^s G_{-1}^{-1}(D)$ is a polynomial matrix in D it follows that

$$\boldsymbol{v}(D)D^s G_{-1}^{-1}(D) = \boldsymbol{u}(D)G(D)D^s G_{-1}^{-1}(D) = \boldsymbol{u}(D)D^s \tag{84}$$

is polynomial, i.e., $\boldsymbol{u}(D)$ has finitely many terms. But $\boldsymbol{u}(D)$ is a formal power series, hence, we conclude that it is polynomial in D. By Lemma 18, $G(D)$ has a polynomial right inverse in D.

d)$\Rightarrow$c): Assume that the abstract state $\boldsymbol{s}_G(D)$ of $G(D)$ is a codeword. That is,
$$\boldsymbol{s}_G(D) = \boldsymbol{u}(D)G(D)Q = \boldsymbol{u}'(D)G(D), \tag{85}$$
where $\boldsymbol{u}(D)$ is polynomial in D^{-1} but without a constant term and $\boldsymbol{u}'(D) \in \boldsymbol{F}_2^b((D))$. Since $\boldsymbol{s}_G(D)$ is a formal power series and $G(D)$ has a polynomial right inverse, it follows that $\boldsymbol{u}'(D)$ is also a formal power series.

Let us use the fact that
$$Q = 1 + P \tag{86}$$
and rewrite (85) as follows:
$$\boldsymbol{s}_G(D) = \boldsymbol{u}(D)G(D) + \boldsymbol{u}(D)G(D)P = \boldsymbol{u}'(D)G(D). \tag{87}$$

Let $G_{-1}^{-1}(D)$ be a right inverse of $G(D)$ whose entries are polynomials in D^{-1}. Then
$$\boldsymbol{u}(D)G(D)G_{-1}^{-1}(D) + \boldsymbol{u}(D)G(D)PG_{-1}^{-1}(D) = \boldsymbol{u}'(D)G(D)G_{-1}^{-1}(D), \tag{88}$$
which can be simplified to
$$\boldsymbol{u}(D) + \boldsymbol{u}(D)G(D)PG_{-1}^{-1}(D) = \boldsymbol{u}'(D). \tag{89}$$

Since $\boldsymbol{u}(D)G(D)P$ is polynomial in D^{-1} without a constant term, it follows that $\boldsymbol{u}(D)G(D)PG_{-1}^{-1}(D)$ is polynomial in D^{-1} without a constant term. Furthermore, $\boldsymbol{u}(D)$ is polynomial in D^{-1} without a constant term and $\boldsymbol{u}'(D)$ is a formal power series. Thus, we conclude that $\boldsymbol{u}'(D) = \boldsymbol{0}$ and, hence, that $\boldsymbol{s}_G(D) = \boldsymbol{0}$. □

The simple minimality test d) is a new result related to Forney's *global invertability test* [7] and to the minimality test of Loeliger and Mittelholzer [8].

Corollary 20 *Every minimal encoding matrix is noncatastrophic.*

Proof It was pointed out by Forney [1] that a convolutional encoding matrix is non-catastrophic if and only if it has a polynomial right pseudo-inverse. Hence, our corollary follows immediately from Theorem 19. □

The following simple example shows that not all basic encoding matrices are minimal.

Example 4 Consider the basic encoding matrix

$$G(D) = \begin{pmatrix} 1+D & D \\ D & 1+D \end{pmatrix}, \tag{90}$$

which has $\mu = 0$ but $\nu = 2$. Clearly, it is not minimal-basic.

The *equivalent* minimal-basic encoding matrix,

$$G_{mb}(D) = \begin{pmatrix} 1 & 0 \\ 1 & 1 \end{pmatrix}, \tag{91}$$

has only one abstract state, viz., $s_{mb} = (0,0)$, and can, of course, be realized without any memory element.

Since $G(D)$ has two abstract states, viz., $s_0 = (0,0)$ and $s_1 = (1,1)$, it is not minimal!

Moreover, $G(D)$ is invertible and its unique inverse is $G(D)$ itself. But

$$G^{-1}(D) = G(D) = \begin{pmatrix} 1+D & D \\ D & 1+D \end{pmatrix} = \begin{pmatrix} \frac{1+D^{-1}}{D^{-1}} & \frac{1}{D^{-1}} \\ \frac{1}{D^{-1}} & \frac{1+D^{-1}}{D^{-1}} \end{pmatrix}, \tag{92}$$

which is not a polynomial matrix in D^{-1}. By Theorem 16 we deduce again that $G(D)$ is not minimal.

Before we state a theorem on when a basic encoding matrix is minimal, we shall prove two lemmas.

Lemma 21 Let $f_1(D), f_2(D), \cdots, f_l D \in \mathbf{F}_2[D]$ with

$$(f_1(D), f_2(D), \cdots, f_l(D)) = 1, \tag{93}$$

where $(f_1(D), f_2(D), \cdots, f_l(D))$ denotes the greatest common divisor of $f_1(D), f_2(D), \cdots, f_l(D)$, and let

$$n = \max(\deg f_1(D), \deg f_2(D), \cdots, \deg f_l(D)). \tag{94}$$

Then for $m \geqslant n D^{-m} f_1(D), D^{-m} f_2(D), \cdots, D^{-m} f_l(D) \in \mathbf{F}_2[D^{-1}]$ and

$$(D^{-m} f_1(D), D^{-m} f_2(D), \cdots, D^{-m} f_l(D)) = D^{-(m-n)}. \quad (95)$$

Proof Let

$$f_i(D) = D^{s_i} g_i(D), \quad i = 1, 2, \cdots, l, \quad (96)$$

where s_i is the start of $f_i(D)$ and $g_i(D) \in \mathbf{F}_2[D]$ is delayfree. From (93) follows

$$\min(s_1, s_2, \cdots, s_l) = 0 \quad (97)$$

and

$$(g_1(D), g_2(D), \cdots, g_l(D)) = 1. \quad (98)$$

For $m \geqslant n$

$$\begin{aligned}
D^{-m} f_i(D) &= D^{-m} D^{s_i} g_i(D) \\
&= D^{-(m - s_i - \deg g_i(D))} \left(D^{-\deg g_i(D)} g_i(D) \right) \\
&= D^{-(m - \deg f_i(D))} \left(D^{-\deg g_i(D)} g_i(D) \right), \\
& \quad i = 1, 2, \cdots, l,
\end{aligned} \quad (99)$$

where the last equality follows from the fact that

$$\deg f_i(D) = s_i + \deg g_i(D), \quad i = 1, 2, \cdots, l. \quad (100)$$

Since $D^{-\deg g_i(D)} g_i(D), i = 1, 2, \cdots, l$, are delayfree it follows from (99) that

$$\begin{aligned}
&(D^{-m} f_1(D), D^{-m} f_2(D), \cdots, D^{-m} f_l(D)) \\
&= D^{-(m-n)} (D^{-\deg g_1(D)} g_1(D), D^{-\deg g_2(D)} g_2(D), \cdots, \\
& \quad D^{-\deg g_l(D)} g_l(D)).
\end{aligned} \quad (101)$$

Clearly,

$$(D^{-\deg g_1(D)} g_1(D), D^{-\deg g_2(D)} g_2(D),$$

$$\ldots, D^{-\deg g_l(D)} g_l(D)) = 1 \tag{102}$$

and the proof is completed. □

Lemma 22 *Let $G(D)$ be a basic encoding matrix and let r and s be the maximum degree of its $b \times b$ minors and $(b-1) \times (b-1)$ minors, respectively. Then the b-th invariant factor of $G(D)$ regarded as a matrix over $\boldsymbol{F}_2(D^{-1})$ is $1/D^{-(r-s)}$.*

Proof Let $G(D) = (g_{ij}(D)), 1 \leqslant i \leqslant b, 1 \leqslant j \leqslant c$, and let $n = \max_{i,j}(\deg g_{ij}(D))$. Write $G(D)$ as a matrix over $\boldsymbol{F}_2(D^{-1})$ as follows:

$$G(D) = \left(\frac{D^{-n} g_{ij}(D)}{D^{-n}}\right)_{i,j} = \frac{1}{D^{-n}} G_{-1}(D), \tag{103}$$

where

$$G_{-1}(D) = (D^{-n} g_{ij}(D))_{i,j} \tag{104}$$

is a matrix of polynomials in D^{-1}.

Since $G(D)$ is basic it follows, by definition, that it has a polynomial right inverse. Hence, it follows from the invariant-factor decomposition (70) that

$$\alpha_1(D) = \alpha_2(D) = \cdots = \alpha_b(D) = 1 \tag{105}$$

(all $\beta_i(D)$ are trivially 1 for a polynomial matrix). Let $\Delta_i(G(D))$ be the greatest common divisor of the $i \times i$ minors of $G(D)$. Since [5]

$$\Delta_i(G(D)) = \alpha_1(D)\alpha_2(D)\cdots\alpha_i(D), \tag{106}$$

we have in particular

$$\Delta_b(G(D)) = \Delta_{b-1}(G(D)) = 1. \tag{107}$$

An $i \times i$ minor of $G_{-1}(D)$ is equal to the corresponding minor of $G(D)$ multiplied by D^{-ni}. Hence, by Lemma 21, we have

$$\Delta_b(G_{-1}(D)) = D^{-(nb-r)} \tag{108}$$

and
$$\Delta_{b-1}(G_{-1}(D)) = D^{-(n(b-1)-s)}. \tag{109}$$

Thus the bth invariant-factor of $G_{-1}(D)$ is [5]
$$\frac{\Delta_b(G_{-1}(D))}{\Delta_{b-1}(G_{-1}(D))} = \frac{D^{-n}}{D^{-(r-s)}}. \tag{110}$$

From (103) and (110), it follows that the bth invariant factor of $G(D)$, regarded as a matrix over $\boldsymbol{F}_2[D^{-1}]$, is
$$\frac{1}{D^{-n}} \cdot \frac{D^{-n}}{D^{-(r-s)}} = \frac{1}{D^{-(r-s)}}. \quad \square$$

We are now ready to prove the following new theorem which was recently formulated by Forney [9].

Theorem 23   *A basic encoding matrix $G(D)$ is minimal if and only if the maximum degree of its $b \times b$ minors is not less than the maximum degree of its $(b-1) \times (b-1)$ minors.*

Proof   From Theorem 19, it follows that a basic encoding matrix $G(D)$ is minimal if and only if it has a polynomial right inverse in D^{-1}. By the invariant factor decomposition $G(D)$ has a polynomial right inverse in D^{-1} if and only if the inverse of its bth invariant factor, regarded as a matrix over $\boldsymbol{F}_2[D^{-1}]$, is a polynomial in D^{-1}. By applying Lemma 22, the theorem follows. $\square$

This theorem follows also from Forney's global invertibility test [7].

VII. Minimal Encoders

We shall now return to our favorite encoding matrix given in Example 2, viz.,
$$G'(D) = \begin{pmatrix} 1+D & D & 1 \\ 1+D^2+D^3 & 1+D+D^2+D^3 & 0 \end{pmatrix}. \tag{111}$$

In Example 3, we showed that $G'(D)$ is not minimal-basic, i.e., $\mu < \nu$. Its controller canonical form (Fig. 7) requires four memory elements but the controller canonical form of an equivalent encoding matrix (Fig. 6) requires only three memory elements. However, $G'(D)$ is a basic encoding matrix and, hence, it has a polynomial right inverse. Furthermore, it has a polynomial right inverse in D^{-1}, viz.,

$$G_{-1}^{-1}(D) = \begin{pmatrix} 1 + D^{-1} + D^{-2} + D^{-3} & D^{-1} \\ 1 + D^{-1} + D^{-3} & D^{-1} \\ D^{-2} + D^{-3} & D^{-1} \end{pmatrix}. \qquad (112)$$

Thus, from Theorem 19, we conclude that $G'(D)$ is indeed a minimal encoding matrix!

Definition A *minimal encoder* is a realization of a minimal encoding matrix $G(D)$ with a minimal number of memory elements over all realizations of $G(D)$.

Theorem 24 (Forney [1]) *The controller canonical form of a minimal-basic encoding matrix is a minimal encoder.*

Proof The proof follows immediately from Corollary 7 and Corollary 17. □

Example 5 The realization shown in Fig. 8 of the minimal encoding matrix $G'(D)$ given in (111) is a minimal encoder. (This realization was obtained by minimizing $G'(D)$ using a standard sequential circuits minimization method.)

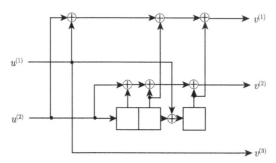

Fig. 8 Minimal encoder for the encoding matrix $G'(D)$ given in (111)

Notice that the minimal realization shown in Fig. 8 is neither in controller canonical nor observer canonical form! This particular minimal encoding matrix does not have a *minimal* controller canonical form, but it has, of course, an *equivalent minimal-basic* encoding matrix whose controller canonical form (Fig. 6) is a minimal encoder for the same convolutional code.

VIII. Systematic Encoders

In a rate $R = b/c$ convolutional code encoded by a systematic encoding matrix, the b information sequences appear among the c output sequences. Without loss of generality, we assume that the first b output sequences are the exact replicas of the b input sequences. Systematic convolutional encoding matrices are simpler to implement, have trivial right inverses, but unless we use rational encoding matrices, i.e., allow feedback in the encoder, they are, as we know (see e.g., [10]), in general, less powerful when used together with maximum likelihood decoding.

A basic encoding matrix has the greatest common divisor of all $b \times b$ minors equal to 1 [1]. Thus, it follows that every basic encoding matrix must have some $b \times b$ submatrix whose determinant is a delayfree polynomial, since otherwise all subdeterminants would be divisible by D. Premultiplication by the inverse of such a submatrix yields an equivalent systematic encoding matrix, possibly rational. Thus, we have the following.

Theorem 25 (Costello [11]) *Every convolutional encoding matrix is equivalent to a systematic rational encoding matrix.*

Example 6 Consider the rate $R = 2/3$ nonsystematic convolutional encoder illustrated in Fig.6. It has the minimal-basic encoding matrix

$$G(D) = \begin{pmatrix} 1+D & D & 1 \\ D^2 & 1 & 1+D+D^2 \end{pmatrix}$$

with $\mu = \nu = 3$. Let $T(D)$ be the matrix consisting of the first two columns of $G(D)$:

$$T(D) = \begin{pmatrix} 1+D & D \\ D^2 & 1 \end{pmatrix}.$$

We have $\det(T(D)) = 1 + D + D^3$, and

$$T^{-1}(D) = \frac{1}{1+D+D^3} \begin{pmatrix} 1 & D \\ D^2 & 1+D \end{pmatrix}.$$

Multiplying $G(D)$ by $T^{-1}(D)$ yields a systematic encoding matrix $G_{\text{sys}}(D)$ equivalent to $G(D)$:

$$\begin{aligned}
G_{\text{sys}}(D) &= T^{-1}(D)G(D) \\
&= \frac{1}{1+D+D^3} \begin{pmatrix} 1 & D \\ D^2 & 1+D \end{pmatrix} \begin{pmatrix} 1+D & D & 1 \\ D^2 & 1 & 1+D+D^2 \end{pmatrix} \\
&= \begin{pmatrix} 1 & 0 & \dfrac{1+D+D^2+D^3}{1+D+D^3} \\ 0 & 1 & \dfrac{1+D^2+D^3}{1+D+D^3} \end{pmatrix}.
\end{aligned}$$

Its observer canonical form requires a linear sequential circuit with feedback and $\mu = 3$ memory elements as shown in Fig. 9.

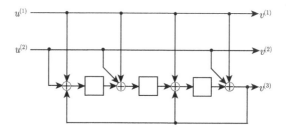

Fig. 9   Observer canonical form of the systematic encoding matrix in Example 6

The systematic encoding matrix in the previous example was realized with the same number of memory elements as the equivalent minimal-basic encoding matrix (Example 3). Hence, it is a minimal encoding matrix. Every systematic encoding matrix can be written (15)

$$G(D) = (I_b R(D)), \tag{113}$$

where I_b is a $b \times b$ identity matrix and $R(D)$ a $b \times (c-b)$ matrix whose entries are rational functions of D. Such a systematic encoding matrix $G(D)$ has a trivial right inverse, viz., the $c \times b$ matrix

$$G^{-1}(D) = \begin{pmatrix} I_b \\ 0 \end{pmatrix}, \qquad (114)$$

which is polynomial in both D and D^{-1}. Hence, it follows from Theorem 19 that this minimality holds in general.

Theorem 26 (Forney [1]) *Every systematic encoding matrix is minimal.*

An Algorithm to Construct a (Minimal) Systematic Encoding Matrix Equivalent to a Given Minimal-Basic Encoding Matrix:

1) Find a $b \times b$ nonzero minor of the minimal-basic encoding matrix $G(D)$ and let it be the determinant of the $b \times b$ submatrix $T(D)$ of $G(D)$.

2) Compute

$$T^{-1}(D)G(D),$$

which is a (minimal) systematic encoding matrix equivalent to $G(D)$.

3) If $T^{-1}(D)G(D) \neq (I_b \ R(D))$, then permute the columns of $T^{-1}(D)G(D)$ and form $G_{\text{sys}}(D) = (I_b \ R(D))$, where $G_{\text{sys}}(D)$ is a (minimal) systematic encoding matrix that encodes a convolutional code which is *equivalent* to the code that $G(D)$ encodes.

Example 7 Consider the rate $R = 2/4$ minimal-basic encoding matrix

$$G(D) = \begin{pmatrix} 1+D & D & 1 & D \\ D & 1 & D & 1+D \end{pmatrix}$$

with $\mu = \nu = 2$. Let

$$T(D) = \begin{pmatrix} 1+D & D \\ D & 1 \end{pmatrix}.$$

Then, we have

$$T^{-1}(D) = \frac{1}{1+D+D^2} \begin{pmatrix} 1 & D \\ D & 1+D \end{pmatrix}$$

and

$$G_{\text{sys}}(D) = T^{-1}(D)G(D)$$

$$= \frac{1}{1+D+D^2} \begin{pmatrix} 1 & D \\ D & 1+D \end{pmatrix} \begin{pmatrix} 1+D & D & 1 & D \\ D & 1 & D & 1+D \end{pmatrix}$$

$$= \begin{pmatrix} 1 & 0 & \dfrac{1+D^2}{1+D+D^2} & \dfrac{D^2}{1+D+D^2} \\ 0 & 1 & \dfrac{D^2}{1+D+D^2} & \dfrac{1}{1+D+D^2} \end{pmatrix}.$$

Clearly, $G_{\text{sys}}(D)$ has neither a minimal controller canonical form nor a minimal observer canonical form but by standard minimization techniques for sequential circuits we obtain the minimal realization shown in Fig. 10.

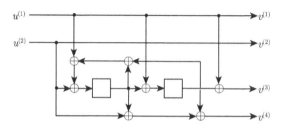

Fig. 10 Minimal realization of the systematic encoding matrix in Example 7

IX. Conclusion

In this semitutorial paper, we have given a summary of some of Forney's previous work along with a few new contributions. Most important among the new results are Theorem 19d), the surprising fact that there exist basic encoding matrices that are minimal but not minimal-basic, the existence of basic encoding matrices that are nonminimal, and a recent result, due to Forney, which states exactly when a basic encoding matrix is minimal (Theorem 23).

X. Comments

Current work on minimal encoders over groups (e.g., Forney and Trott [12], Loeliger and Mittelholzer [8]) constructs canonical minimal encoders from a set of shortest linearly independent code sequences ("trellis-oriented generators"), a type of construction that may have been first published by Roos [13]; see also Piret [14]. In Forney [7], the approach of [1] is extended to generalized minimal encoders that are in controller canonical form but not necessarily polynomial.

Acknowledgment

The authors' debt to D. Forney is both obvious and gratefully acknowledged. The authors are also grateful to J. Massey for his interest and encouragement over the years. Without their stimulation, the authors would not have reached the view of convolutional codes that is presented in this paper. D. Forney's detailed comments on the manuscript have been of great value.

References

[1] G. D. Forney, Jr., "Convolutional codes I: Algebraic structure," *IEEE Trans. Inform. Theory*, vol. IT-16, pp. 720-738, 1970.

[2] G. D. Forney, Jr., "Correction to 'Convolutional codes I: Algebraic structure,'" *IEEE Trans. Inform. Theory*, vol. IT-17, pp. 360, 1971.

[3] G. D. Forney, Jr., "Structural analyses of convolutional codes via dual codes," *IEEE Trans. Inform. Theory*, vol. IT-19, pp. 512-518, 1973.

[4] J. L. Massey and M. K. Sain, "Inverses of linear sequential circuits," *IEEE Trans. Comput*, vol. C-17, pp. 330-337, 1968.

[5] N. Jacobson, *Basic Algebra I*. San Fransisco, CA: Freeman, 1974.

[6] T. Kailath, *Linear systems*. Englewood Cliffs, NJ: Prentice Hall, 1980.

[7] G. D. Forney, Jr., "Algebraic structure of convolutional codes, and algebraic system theory," in *Mathematical System Theory*, A. C. Antoulas, Ed. Berlin: Springer-Verlag, 1991, pp. 527-558.

[8] H.-A. Loeliger and T. Mittelholzer, "Convolutional codes over groups," submitted to *IEEE Trans. Inform. Theory*, 1992.

[9] G.D. Forney, Jr., Private communication, Aug. 1991.

[10] A. J. Viterbi and J. K. Omura, *Principles of Digital Communication and Coding*. New York: McGraw-Hill, 1979.

[11] D. J. Costello, Jr., "Construction of convolutional codes for sequential decoding," Rep. EE-692, Univ. of Notre Dame, Notre Dame, IN, Aug. 1969.

[12] G. D. Forney, Jr. and M. D. Trott, "The dynamics of linear codes over groups: state spaces, trellis diagrams and canonical encoders," submitted to *IEEE Trans. Inform. Theory*, 1992.

[13] C. Roos, "On the structure of convolutional and cyclic convolutional codes," *IEEE Trans. Inform. Theory*, vol. IT-25, pp. 676-683, 1979.

[14] P. Piret, *Convolutional Codes: An Algebraic Approach*. Cambridge, MA: MIT Press, 1988.

Representations of Forms by Forms in a Finite Field

Zhexian Wan

Institute of Systems Science, Chinese Academy of Sciences, Beijing 100080; China;
and Department of Information Theory, Lund University,
Box 118, S-221 00 Lund, Sweden
E-mail: wan@dit.lth.se

Communicated by James W. P. Hirschfeld

Received April 1, 1994

Abstract Let $\mathscr{F}_q$ be a finite field with q elements, where q is a prime power which may be either odd or even. Representations of a form by another form of the same type in $\mathscr{F}_q$ are studied and formulas of the number of representations are obtained. Forms considered in this paper include bilinear forms, alternate bilinear forms, hermitian forms, quadratic forms, and symmetric bilinear forms. © 1995 Academic Press. Inc.

1. Introduction

Carlitz studied the representations of a form by another one of the same type in a finite field of odd characteristic in 1954. He obtained formulas for the number of representations of a quadratic (or skew-symmetric) form by another quadratic (or skew-symmetric) form in a finite field of odd characteristic (cf. [1,2]) and, together with Hodges, formulas for the number of representations of an hermitian form by another hermitian form also in a finite field of odd characteristic (cf. [3]).

In the present paper the above problem will be studied in its full generality.

On the one hand, the restriction that the finite field is of odd characteristic is removed. On the other hand, in addition to the quadratic, skewsymmetric, and hermitian forms studied by Carlitz, the bilinear, alternate bilinear, and symmetric bilinear forms are also studied.

Throughout this paper let $\mathscr{F}_q$ denote the finite field with q elements, where q is a prime power. Unless otherwise stated, q may be odd or even.

This paper is organized as follows. In Section 2 representations of a bilinear form by another bilinear form in $\mathscr{F}_q$ are studied and formulas for the number of representations are deduced. The corresponding problems for alternate bilinear forms in $\mathscr{F}_q$, for hermitian forms in $\mathscr{F}_q$ where q is a perfect square, for quadratic forms in $\mathscr{F}_q$ where q is odd, for quadratic form in $\mathscr{F}_q$ where q is even, and for symmetric bilinear forms in $\mathscr{F}_q$ where q is even are studied in Sections 3, 4, 5, 6 and 7, respectively.

2. Representations of Bilinear Forms

Let A be an $m \times n$ matrix over $\mathscr{F}_q$ and B be a $t \times u$ matrix over $\mathscr{F}_q$. Denote by $n^{(b)}_{m \times t, n \times u}(A, B)$ the number of pairs of $m \times t$ and $n \times u$ matrices (X, Y) over $\mathscr{F}_q$ such that

$$^t X A Y = B. \tag{1}$$

Our purpose is to deduce formulas for $n^{(b)}_{m \times t, n \times u}(A, B)$. Clearly there is no loss of generality in assuming that one or both of A and B are in their normal forms under equivalence transformations. We recall that an $m \times n$ matrix A is equivalent to one of the normal forms

$$\begin{pmatrix} I^{(r)} & \\ & 0^{(m-r, n-r)} \end{pmatrix}, \quad r = 0, 1, 2, \cdots, \min\{m, n\}, \tag{2}$$

where r is the rank of A. Here and later we adopt the convention that zeros in matrices are sometimes omitted; i.e., the blanks in matrices replace the omitted zeros if it is clear from the context.

Proposition 1 *Let A be an $m \times n$ matrix of rank r over $\mathscr{F}_q$, where $0 \leqslant r \leqslant \min\{m,n\}$, and B be a $t \times u$ matrix over $\mathscr{F}_q$. Then*

$$n^{(b)}_{m \times t, n \times u}(A, B) = q^{t(m-r)+u(n-r)} n^{(b)}_{r \times t, r \times u}(I^{(r)}, B). \tag{3}$$

Proof We can assume that A is of the form (2). Let X be an $m \times t$ matrix and Y be an $n \times u$ matrix satisfying (1). Write X and Y in the following block forms:

$$X = \begin{pmatrix} X_1 \\ X_2 \end{pmatrix} \begin{matrix} r \\ m-r \end{matrix} \overset{t}{} \tag{4}$$

and

$$Y = \begin{pmatrix} Y_1 \\ Y_2 \end{pmatrix} \begin{matrix} r \\ n-r \end{matrix} \overset{u}{} ; \tag{5}$$

then

$${}^tXAY = {}^tX_1 I^{(r)} Y_1 = B. \tag{6}$$

Therefore X_2 and Y_2 can be arbitrary, and X_1 and Y_1 satisfy ${}^tX_1 I^{(r)} Y_1 = B$. Hence (3) follows. □

Thus our problem is reduced to computing $n^{(b)}_{r \times t, r \times u}(I^{(r)}, B)$.

Proposition 2 *Let B be a $t \times u$ matrix of rank s over $\mathscr{F}_q$, where $0 \leqslant s \leqslant \min\{t,u\}$. Then*

$$n^{(b)}_{r \times t, r \times u}(I^{(r)}, B)$$
$$= \begin{cases} 0 & \text{if } r < s, \\ q^{s(r-s)+(1/2)s(s-1)} \prod_{i=r-s+1}^{r} (q^i - 1) n^{(b)}_{(r-s) \times (t-s), (r-s) \times (u-s)}(I^{(r-s)}, 0^{(t-s, u-s)}) \\ \text{if } r \geqslant s. \end{cases} \tag{7}$$

Proof We can assume that

$$B = \begin{pmatrix} I^{(s)} & \\ & 0^{(t-s, u-s)} \end{pmatrix}. \tag{8}$$

Let X be an $r \times t$ matrix and Y be an $r \times u$ matrix such that

$$^tX I^{(r)} Y = B. \tag{9}$$

Clearly, rank $^tX I^{(r)} Y \leqslant r$. Thus, if $r < s$, then $n^{(b)}_{r \times t, r \times u}(I^{(r)}, B) = 0$.

Now consider the case $r \geqslant s$. Write X and Y in the following block forms:

$$X = \begin{pmatrix} \overset{s}{X_1} & \overset{t-s}{X_2} \end{pmatrix} r \tag{10}$$

and

$$Y = \begin{pmatrix} \overset{s}{Y_1} & \overset{u-s}{Y_2} \end{pmatrix} r \; ; \tag{11}$$

then (9) can be written as

$$^tX_1 Y_1 = I^{(s)}, \tag{12}$$
$$^tX_1 Y_2 = 0, \tag{13}$$
$$^tX_2 Y_1 = 0, \tag{14}$$

and

$$^tX_2 Y_2 = 0. \tag{15}$$

From (12) we deduce that both X_1 and Y_1 are of rank s. We assert that for any $r \times s$ matrix X_1 of rank s, the number of $r \times (t-s)$ matrices X_2, $r \times s$ matrices Y_1, and $r \times (u-s)$ matrices Y_2 for which (12)–(15) hold is independent of the particular choice of X_1. Let X_1 and X_1^* be two $r \times s$ matrices of rank s; then there is an $r \times r$ nonsingular matrix P such that

$$X_1^* = P X_1. \tag{16}$$

Assume that X_2, Y_1 and Y_2 are $r \times (t-s), r \times s$, and $r \times (u-s)$ matrices, respectively, for which (12)–(15) hold. Let

$$X_2^* = P X_2, \quad Y_1^* = {}^tP^{-1} Y_1, \quad Y_2^* = {}^tP^{-1} Y_2; \tag{17}$$

then X_1^*, X_2^*, Y_1^*, and Y_2^* also satisfy (12)–(15). This proves our assertion.

The number of $r \times s$ matrices of rank s over $\mathscr{F}_q$ is equal to

$$q^{(1/2)s(s-1)} \prod_{i=r-s+1}^{r} (q^i - 1), \tag{18}$$

(cf. Lemma 1.5 of [5]). Given any $r \times s$ matrix X_1 of rank s, we have to compute the number of $r \times (t-s)$ matrices X_2, $r \times s$ matrices Y_1, and $r \times (u-s)$ matrices Y_2 such that (12)–(15) hold. We may choose

$$X_1 = \begin{pmatrix} I^{(s)} \\ 0^{(r-s,s)} \end{pmatrix}. \tag{19}$$

Then from (12) and (13) we deduce

$$Y_1 = \begin{pmatrix} I^{(s)} \\ Y_{12} \end{pmatrix} \begin{matrix} s \\ r-s \end{matrix} \tag{20}$$

and

$$Y_2 = \begin{pmatrix} 0 \\ Y_{22} \end{pmatrix} \begin{matrix} s \\ r-s \end{matrix} \tag{21}$$

Write

$$X_2 = \begin{pmatrix} X_{21} \\ X_{22} \end{pmatrix} \begin{matrix} s \\ r-s \end{matrix} ; \tag{22}$$

then from (14) and (15) we deduce

$$^tX_{21} + {}^tX_{22}Y_{12} = 0 \tag{23}$$

and

$$^tX_{22}Y_{22} = 0, \tag{24}$$

respectively. The number of pairs of $(r-s) \times (t-s)$ and $(r-s) \times (u-s)$ matrices (X_{22}, Y_{22}) satisfying (24) is denoted by

$$n^{(b)}_{(r-s) \times (t-s),(r-s) \times (u-s)}(I^{(r-s)}, 0^{(t-s,u-s)}). \tag{25}$$

Given any pair of $(r-s) \times (t-s)$ and $(r-s) \times (u-s)$ matrices (X_{22}, Y_{22}) satisfying (24) and any $(r-s) \times s$ matrix Y_{12}, there is a uniquely determined $s \times (t-s)$ matrix X_{21} satisfying (23). Therefore, given any $r \times s$ matrix X_1 of rank s, the number of $r \times (t-s)$ matrices X_2, $r \times s$ matrices Y_1, and $r \times (u-s)$ matrices Y_2 for which (12)—(15) hold is equal to

$$q^{s(r-s)} n^{(b)}_{(r-s)\times(t-s),(r-s)\times(u-s)}(I^{(r-s)}, 0^{(t-s,u-s)}). \tag{26}$$

Multiplying (26) by (18), we obtain the second formula of (7). □

Corollary 3 *Let $r \geqslant s$. Then*

$$n^{(b)}_{r\times s, r\times s}(I^{(r)}, I^{(s)}) = q^{s(r-s)+(1/2)s(s-1)} \prod_{i=r-s+1}^{r} (q^i - 1). \tag{27}$$

Proposition 4

$$n^{(b)}_{m\times t, m\times n}(I^{(m)}, 0^{(t,u)})$$

$$= \sum_{k=0}^{\min\{m,u\}} q^{t(m-k)+(1/2)k(k-1)} \frac{\prod_{i=m-k+1}^{m}(q^i-1) \prod_{i=u-k+1}^{u}(q^i-1)}{\prod_{i=1}^{k}(q^i-1)}. \tag{28}$$

Proof $n^{(b)}_{m\times t, m\times u}(I^{(m)}, 0^{(t,u)})$ is the number of pairs of $m \times t$ and $m \times u$ matrices (X, Y) such that

$${}^t XY = 0^{(t,u)}. \tag{29}$$

Let Y be an $m \times u$ matrix. Denote by U the subspace spanned by the columns of Y. Let $\dim U = k$, i.e., rank $Y = k$; then $k \leqslant \min\{m, u\}$. Let $u_1, u_2, \cdots, u_k$ be a basis of U; then there is a uniquely determined $k \times u$ matrix T of rank k such that

$$Y = (u_1, u_2, \cdots, u_k)T. \tag{30}$$

Hence the number of $m \times u$ matrices of rank k is equal to

$$\begin{bmatrix} m \\ k \end{bmatrix}_q q^{(1/2)k(k-1)} \prod_{i=u-k+1}^{u} (q^i - 1), \tag{31}$$

where

$$\begin{bmatrix} m \\ k \end{bmatrix}_q = \frac{\prod_{i=m-k+1}^{m}(q^i-1)}{\prod_{i=1}^{k}(q^i-1)} \qquad (32)$$

is the number of k-dimensional subspaces of the m-dimensional column vector space $\mathscr{F}_q^{(m)}$. Given any $m \times u$ matrix Y of rank k, the number of $m \times t$ matrices X satisfying (29) is equal to $q^{t(m-k)}$. Hence (28) follows. $\square$

Corollary 5

$$\sum_{k=0}^{\min(m,u)} q^{t(m-k)+(1/2)k(k-1)} \frac{\prod_{i=m-k+1}^{m}(q^i-1) \prod_{i=u-k+1}^{u}(q^i-1)}{\prod_{i=1}^{k}(q^i-1)}$$

$$= \sum_{k=0}^{\min\{m,t\}} q^{u(m-k)+(1/2)k(k-1)} \frac{\prod_{i=m-k+1}^{m}(q^i-1) \prod_{i=t-k+1}^{t}(q^i-1)}{\prod_{i=1}^{k}(q^i-1)}. \qquad (33)$$

From these propositions it is not difficult to deduce a closed formula for $n_{m \times t, n \times u}^{(b)}(A, B)$, where A is an $m \times n$ matrix and B is a $t \times u$ matrix.

3. Representations of Alternate Forms

Let A be an $m \times m$ alternate matrix over $\mathscr{F}_q$ and B be a $t \times t$ alternate matrix over $\mathscr{F}_q$. Denote by $n_{m \times t}^{(u)}(A, B)$ the number of $m \times t$ matrices X over $\mathscr{F}_q$ such that

$${}^tXAX = B. \qquad (34)$$

Our purpose is to deduce formulas for $n_{m \times t}^{(u)}(A, B)$. When q is odd, such formulas were obtained previously by Carlitz [2].

Clearly, there is no loss of generality in assuming that one or both of A and B are in their normal forms under cogredience transformations. We recall that an $m \times m$ alternate matrix A is cogredient to one of the normal forms

$$\begin{pmatrix} 0 & I^{(\nu)} & \\ -I^{(\nu)} & 0 & \\ & & 0^{(m-2\nu)} \end{pmatrix}, \quad \nu = 0, 1, 2, \cdots, \left[\frac{m}{2}\right], \tag{35}$$

where ν is called the *index* of A, (cf. Theorem 3.1 of [5]). Let

$$K_{2\nu} = \begin{pmatrix} 0 & I^{(\nu)} \\ -I^{(\nu)} & 0 \end{pmatrix}. \tag{36}$$

Then we have

Proposition 6 Let A be an $m \times m$ alternate matrix of index ν over $\mathscr{F}_q$, where $2\nu \leqslant m$, and B be a $t \times t$ alternate matrix over $\mathscr{F}_q$. Then

$$n_{m \times t}^{(a)}(A, B) = q^{t(m-2\nu)} n_{2\nu \times t}^{(a)}(K_{2\nu}, B). \tag{37}$$

Proof We can assume that A is of the form (35). Let X be an $m \times t$ matrix over $\mathscr{F}_q$ satisfying (34). Write X in the block form

$$X = \begin{pmatrix} X_1 \\ X_2 \end{pmatrix} \begin{matrix} 2\nu \\ m-2\nu \end{matrix} ; \tag{38}$$

then

$$^tXAX = {}^tX_1 K_{2\nu} X_1 = B. \tag{39}$$

Therefore X_2 can be any $(m-2\nu) \times t$ matrix and X_1 is a $2\nu \times t$ matrix satisfying $^tX_1 K_{2\nu} X_1 = B$. Hence (37) follows. □

Proposition 7 Let B be a $t \times t$ alternate matrix of index ν_1 over $\mathscr{F}_q$, where $2\nu_1 \leqslant t$. Then

$$n_{2\nu \times t}^{(a)}(K_{2\nu}, B)$$
$$= \begin{cases} 0 & \text{if } \nu_1 > \nu, \\ n_{2\nu \times 2\nu_1}^{(a)}(K_{2\nu}, K_{2\nu_1}) n_{2(\nu-\nu_1) \times (t-2\nu_1)}^{(a)}(K_{2(\nu-\nu_1)}, 0^{(t-2\nu_1)}) & \text{if } \nu_1 \leqslant \nu. \end{cases} \tag{40}$$

Proof The reduction formula (40) is due to Carlitz (cf. formula (7.2) of [2]). For completeness, we shall give a simpler proof. We can assume that

$$B = \begin{pmatrix} K_{2\nu_1} & \\ & 0^{(t-2\nu_1)} \end{pmatrix}. \tag{41}$$

Let X be a $2\nu \times t$ matrix over $\mathscr{F}_q$ satisfying

$$^tXK_{2\nu}X = B. \tag{42}$$

Write X in the block form

$$X = \begin{pmatrix} \overset{2\nu_1}{X_1} & \overset{t-2\nu_1}{X_2} \end{pmatrix} 2\nu \,. \tag{43}$$

From (42) we deduce

$$^tX_1K_{2\nu}X_1 = K_{2\nu_1}, \tag{44}$$

$$^tX_1K_{2\nu}X_2 = 0^{(2\nu_1, t-2\nu_1)}, \tag{45}$$

and

$$^tX_2K_{2\nu}X_2 = 0^{(t-2\nu_1)}. \tag{46}$$

If $\nu_1 > \nu$, then there is no X_1 satisfying (44) and, hence, $n_{2\nu \times t}^{(a)}(K_{2\nu}, B) = 0$.

Now consider the case $\nu_1 \leqslant \nu$. For any $2\nu_1 \times 2\nu$ matrix X_1 satisfying (44), clearly, we have rank $X_1 = 2\nu_1$. The $2\nu_1$ columns of X_1 span a subspace of dimension $2\nu_1$, which will also be denoted by X_1 if no ambiguity arises. Define

$$X_1^\perp = \{v \in \mathscr{F}_q^{(2\nu)} | {}^tX_1K_{2\nu}v = 0\}, \tag{47}$$

where $\mathscr{F}_q^{(2\nu)}$ is the 2ν-dimensional column vector space over $\mathscr{F}_q$. Then $X_1^\perp$ is a subspace of dimension $2(\nu - \nu_1)$. Clearly, ${}^tX_1KX_1^\perp = 0$. By (44) $X_1 \cap X_1^\perp = \{0\}$. Denote any $2\nu \times 2(\nu - \nu_1)$ matrix over $\mathscr{F}_q$ whose columns form a basis of $X_1^\perp$ also by $X_1^\perp$; then $(X_1 \ X_1^\perp)$ is a $2\nu \times 2\nu$ nonsingular matrix and

$$^t(X_1 \ X_1^\perp)K_{2\nu}(X_1 \ X_1^\perp) = \begin{pmatrix} K_{2\nu_1} & \\ & {}^tX_1^\perp K_{2\nu}X_1^\perp \end{pmatrix}. \tag{48}$$

Therefore ${}^tX_1^\perp K_{2\nu}X_1^\perp$ is a $2(\nu - \nu_1) \times 2(\nu - \nu_1)$ nonsingular alternate matrix. We can assume that

$$^tX_1^\perp K_{2\nu}X_1^\perp = K_{2(\nu-\nu_1)}. \tag{49}$$

It is clear that for any $2(\nu - \nu_1) \times (t - 2\nu_1)$ matrix T satisfying

$$^tT K_{2(\nu-\nu_1)} T = 0, \tag{50}$$

the $2\nu \times t$ matrix

$$(X_1 \ \ X_1^\perp T) \tag{51}$$

satisfies (42). Conversely, for any $2\nu \times t$ matrix (43) satisfying (42), there is a uniquely determined $2(\nu - \nu_1) \times (t - 2\nu_1)$ matrix T such that $X_2 = X_1^\perp T$ and T satisfies (50). Hence the second formula of (40) follows. □

Proposition 8 *Let $\nu_1 \leqslant \nu$. Then*

$$n^{(a)}_{2\nu \times 2\nu_1}(K_{2\nu}, K_{2\nu_1}) = q^{\nu_1(2\nu-\nu_1)} \prod_{i=\nu-\nu_1+1}^{\nu} (q^{2i} - 1). \tag{52}$$

This is Theorem 1 of [2]; see also Lemma 3.15 of [5].

Proposition 9

$$n^{(a)}_{2\nu \times t}(K_{2\nu}, 0^{(t)}) = \sum_{k=0}^{\min\{t,\nu\}} q^{(1/2)k(k-1)} \frac{\prod_{i=t-k+1}^{t}(q^i - 1) \prod_{i=\nu-k+1}^{\nu}(q^{2i}-1)}{\prod_{i=1}^{k}(q^i - 1)}. \tag{53}$$

Proof Let X be a $2\nu \times t$ matrix over $\mathscr{F}_q$ such that

$$^t X K_{2\nu} X = 0. \tag{54}$$

Assume that the columns of X span a subspace U of dimension k. Then $k \leqslant t$ and U is a k-dimensional totally isotropic subspace with respect to $K_{2\nu}$. Hence $k \leqslant \nu$. Let $u_1, u_2, \cdots, u_k$ be a basis of U. Then there is a uniquely determined $k \times t$ matrix T of rank k such that

$$X = (u_1 \ u_2 \cdots u_k) T. \tag{55}$$

Conversely, for any $k \times t$ matrix T of rank k, define X by (55); then X satisfies (54) and the columns of X span U. By Corollary 3.19 of [5] the number of

k-dimensional totally isotropic subspaces with respect to $K_{2\nu}$ is equal to

$$\frac{\prod_{i=\nu-k+1}^{\nu}(q^{2i}-1)}{\prod_{i=1}^{k}(q^i-1)}. \tag{56}$$

By Lemma 1.5 of [5] the number of $k \times t$ matrices of rank k is equal to

$$q^{(1/2)k(k-1)} \prod_{i=t-k+1}^{t}(q^i-1). \tag{57}$$

Hence we have (53). □

Corollary 10

$$n_{2\nu \times t}^{(a)}(K_{2\nu}, 0^{(t)}) = q^{-(1/2)t(t-1)+2\nu t} {}_2\Phi_0[q^{-(1/2)t}, q^{-(1/2)(t-1)}; q^{\nu+1}]', \tag{58}$$

where

$${}_2\Phi_0[a, b; z] = \sum_{k=0}^{z} \frac{(a)_k (b)_k}{(q)_k} z^k, \tag{59}$$

$$(a)_k = (1-a)(1-qa)\cdots(1-q^{k-1}a) \quad if \ k \geqslant 1, (a)_0 = 1, \tag{60}$$

and the prime indicates that q is to be replaced by q^{-2} in the function ${}_2\Phi_0$.

Proof When q is odd, (58) is Theorem 4 of [2]. Therefore when q is odd, we have

$$q^{-(1/2)t(t-1)+2\nu t} {}_2\Phi_0[q^{-(1/2)t}, q^{-(1/2)(t-1)}; q^{\nu+1}]'$$
$$= \sum_{k=0}^{\min\{t,\nu\}} q^{(1/2)k(k-1)} \frac{\prod_{i=t-k+1}^{t}(q^i-1) \prod_{i=\nu-k+1}^{\nu}(q^{2i}-1)}{\prod_{i=1}^{k}(q^i-1)}. \tag{61}$$

Since there are infinitely many odd prime power q, (61) is an identity in q. Therefore (58) holds for any prime power q. □

4. Representations of Hermitian Forms

In this section we assume that q is a perfect square, say $q = q_0^2$, where q_0 is a prime power. Let

$$- : a \to \bar{a} = a^{q_0} \tag{62}$$

be the involution of $\mathscr{F}_q$, whose fixed field is $\mathscr{F}_{q_0}$. A square matrix A over $\mathscr{F}_q$ is called a *hermitian* matrix if

$${}^t\bar{A} = A. \tag{63}$$

Now let A be an $m \times m$ hermitian matrix over $\mathscr{F}_q$ and B be a $t \times t$ hermitian matrix over $\mathscr{F}_q$. Denote by $n_{m \times t}^{(h)}(A, B)$ the number of $m \times t$ matrices X over $\mathscr{F}_q$ such that

$$ {}^t\bar{X} A X = B. \tag{64}$$

When q is odd, formulas for $n_{m \times t}^{(h)}(A, B)$ were obtained previously by Carlitz and Hodges [3]. Now we shall deduce formulas for $n_{m \times t}^{(h)}(A, B)$ for any q.

As in Section 3 there is no loss of generality in assuming that one or both of A and B are in their normal forms under cogredience transformations. We recall that an $m \times m$ hermitian matrix A over $\mathscr{F}_q$ is cogredient to one of the normal forms

$$\begin{pmatrix} I^{(r)} & \\ & 0^{(m-r)} \end{pmatrix}, \quad r = 0, 1, 2, \cdots, m, \tag{65}$$

where r is the rank of A (cf. Theorem 5.2 of [5]). Parallel to Propositions 6-9 and Corollary 10 of Section 3 we have the following propositions whose proofs are the same as the corresponding ones in Section 3 and will be omitted.

Proposition 11 *Let A be an $m \times m$ hermitian matrix of rank r over $\mathscr{F}_q$, where $r \leqslant m$, and B be a $t \times t$ hermitian matrix over $\mathscr{F}_q$. Then*

$$n_{m \times t}^{(h)}(A, B) = q^{t(m-r)} n_{r \times t}^{(h)}(I^{(r)}, B). \tag{66}$$

Proposition 12 *Let B be a $t \times t$ hermitian matrix of rank r_1 over $\mathscr{F}_q$, where $r_1 \leqslant t$. Then*

$$n_{r \times t}^{(h)}(I^{(r)}, B)$$

$$= \begin{cases} 0 & if\ r_1 > r, \\ n^{(h)}_{r \times r_1}(I^{(r)}, I^{(r_1)}) n^{(h)}_{(r-r_1) \times (t-r_1)}(I^{(r-r_1)}, 0^{(t-r_1)}) & if\ r_1 \leqslant r. \end{cases} \quad (67)$$

Proposition 13 *Let $r_1 \leqslant r$. Then*

$$n^{(h)}_{r \times r_1}(I^{(r)}, I^{(r_1)}) = q_0^{r_1 r - (1/2) r_1 (r_1 + 1)} \prod_{i=r-r_1+1}^{r} (q_0^i - (-1)^i). \quad (68)$$

This is Lemma 5.15 of [5].

Both Propositions 12 and 13 are contained in Theorem 1 of [3]; conversely, from Propositions 12 and 13 we can deduce immediately Theorem 1 of [3].

Proposition 14

$$n^{(h)}_{r \times t}(I^{(r)}, 0^{(t)}) = \sum_{k=0}^{\min\{t, [r/2]\}} q_0^{k(k-1)} \frac{\prod_{i=t-k+1}^{t}(q_0^{2i} - 1) \prod_{i=r-2k+1}^{r}(q_0^i - (-1)^i)}{\prod_{i=1}^{k}(q_0^{2i} - 1)}. \quad (69)$$

It should be remarked that in the proof of Proposition 14 the formula of the number of k-dimensional totally isotropic subspaces with respect to $I^{(r)}$, which is given by Corollary 5.20 of [5], is used.

Corollary 15

$$n^{(h)}_{r \times t}(I^{(r)}, 0^{(t)}) = q_0^{2rt - t^2} + \sum_{k=1}^{t}(-1)^{kr} q_0^{2rt - t^2 - rk + (1/2)k(k-1)} \prod_{i=1}^{k} \frac{q_0^{2(t-k+i)} - 1}{q_0^i - (-1)^i}. \quad (70)$$

When q is odd, Corollary 15 was proved previously by Carlitz and Hodges [3].

5. Representations of Quadratic Forms (Odd Characteristic Case)

In this section we assume that q is odd. Let A be an $m \times m$ symmetric matrix and B be a $t \times t$ symmetric matrix, both over $\mathscr{F}_q$. Denote by $n^{(q)}_{m \times t}(A, B)$ the number of $m \times t$ matrix X over $\mathscr{F}_q$ such that

$$^t X A X = B. \quad (71)$$

Formulas for $n^{(q)}_{m\times t}(A, B)$ were obtained previously by Carlitz [2]. Now we shall deduce another set of formulas for $n^{(q)}_{m\times t}(A, B)$ following the same procedure as was used in Section 3.

As in Section 3 there is no loss of generality in assuming that one or both of A and B are in their normal forms under cogredience transformations.

Choose a fixed nonsquare element z in $\mathscr{F}_q$. Let

$$S_{2\nu} = \begin{pmatrix} 0 & I^{(\nu)} \\ I^{(\nu)} & 0 \end{pmatrix}, \tag{72}$$

$$S_{2\nu+1,1} = \begin{pmatrix} 0 & I^{(\nu)} & \\ I^{(\nu)} & 0 & \\ & & 1 \end{pmatrix}, \tag{73}$$

$$S_{2\nu+1,z} = \begin{pmatrix} 0 & I^{(\nu)} & \\ I^{(\nu)} & 0 & \\ & & z \end{pmatrix}, \tag{74}$$

and

$$S_{2\nu+2,\left(\begin{smallmatrix}1 & \\ & -z\end{smallmatrix}\right)} = \begin{pmatrix} 0 & I^{(\nu)} & & \\ I^{(\nu)} & 0 & & \\ & & 1 & \\ & & & -z \end{pmatrix}. \tag{75}$$

In order to cover these four cases we introduce the notation $S_{2\nu+\delta,\Delta}$, where $\delta = 0, 1,$ or 2 and

$$\Delta = \begin{cases} \phi & \text{when } \delta = 0, \\ 1 \text{ or } z & \text{when } \delta = 1, \\ \begin{pmatrix} 1 & \\ & -z \end{pmatrix} & \text{when } \delta = 2. \end{cases} \tag{76}$$

ν is called the *index* of $S_{2\nu+\delta,\Delta}$ and Δ is called the *definite part* of $S_{2\nu+\delta,\Delta}$.

We recall that an $m \times m$ symmetric matrix A over $\mathscr{F}_q$ is cogredient to one of the normal forms

$$\begin{pmatrix} S_{2\nu+\delta,\Delta} & \\ & 0^{(m-2\nu-\delta)} \end{pmatrix}, \quad \nu = 0, 1, 2, \cdots, \left[\frac{m-\delta}{2}\right], \tag{77}$$

where $\delta = 0, 1$, or 2 and Δ is given by (76) (cf. Theorem 1.25 of [5]). Clearly, $2\nu + \delta$ is the rank of A. We called ν the *index* of A and Δ the *definite part* of A.

Parallel to Propositions 6–9 of Section 3 we have the following propositions whose proofs are omitted.

Proposition 16 Let q be odd. Let A be an $m \times m$ symmetric matrix and B be a $t \times t$ symmetric matrix, both over $\mathscr{F}_q$. Assume that A is of index ν, rank $2\nu + \delta$, where $\delta = 0, 1$, or 2 and $2\nu + \delta \leqslant m$, and definite part Δ, which is given by (76). Then

$$n^{(q)}_{m \times t}(A, B) = q^{t(m-2\nu-\delta)} n^{(q)}_{(2\nu+\delta) \times t}(S_{2\nu+\delta, \Delta}, B). \tag{78}$$

Proposition 17 Let q be odd. Let B be a $t \times t$ symmetric matrix over $\mathscr{F}_q$ and assume that B is of index ν_1, rank $2\nu_1 + \delta_1$, where $\delta_1 = 0, 1$, or 2 and $2\nu_1 + \delta_1 \leqslant t$, and definite part Δ_1, where

$$\Delta_1 = \begin{cases} \phi & \text{when } \delta_1 = 0, \\ 1 \text{ or } z & \text{when } \delta_1 = 1, \\ \begin{pmatrix} 1 & \\ & -z \end{pmatrix} & \text{when } \delta_1 = 2. \end{cases} \tag{79}$$

(i) *If the condition*

$$2\nu_1 + \delta_1 \leqslant t \\ \leqslant \begin{cases} \nu + \nu_1 + \min\{\delta, \delta_1\}, & \text{when } \delta_1 \neq \delta, \text{ or } \delta_1 = \delta \text{ and } \Delta_1 = \Delta, \\ \nu + \nu_1, & \text{when } \delta_1 = \delta = 1 \text{ and } \Delta_1 \neq \Delta, \end{cases} \tag{80}$$

does not hold, then

$$n^{(q)}_{(2\nu+\delta) \times t}(S_{2\nu+\delta, \Delta}, B) = 0. \tag{81}$$

(ii) *If condition* (80) *holds, then there exist uniquely determined* ν_2, δ_2, *and* Δ_2, *where* $\delta_2 = 0, 1$, *or* 2 *and*

$$\Delta_2 = \begin{cases} \phi & \text{when } \delta_2 = 0, \\ 1 \text{ or } z & \text{when } \delta_2 = 1, \\ \begin{pmatrix} 1 & \\ & -z \end{pmatrix} & \text{when } \delta_2 = 2, \end{cases} \tag{82}$$

such that

$$\begin{pmatrix} S_{2\nu_1+\delta_1,\Delta_1} & \\ & S_{2\nu_2+\delta_2,\Delta_2} \end{pmatrix} \quad (83)$$

is cogredient to $S_{2\nu+\delta,\Delta}$ and

$$n^{(q)}_{(2\nu+\delta)\times t}(S_{2\nu+\delta,\Delta}, B) = n^{(q)}_{(2\nu+\delta)\times(2\nu_1+\delta_1)}(S_{2\nu+\delta,\Delta}, S_{2\nu_1+\delta_1,\Delta_1})$$
$$\times n^{(q)}_{(2\nu_2+\delta_2)\times(t-2\nu_1-\delta_1)}(S_{2\nu_2+\delta_2,\Delta_2}, 0). \quad (84)$$

We remark that for condition (80), cf. Theorem 6.3 of [5], and that when ν, δ, Δ and ν_1, δ_1, Δ_1, are given and condition (80) holds, the uniquely determined ν_2, δ_2, Δ_2 such that (83) is cogredient to $S_{2\nu+\delta,\Delta}$ are given by Corollary 6.6 of [5].

Proposition 18 *Let q be odd.*

(i) If $\nu \geqslant \nu_1$, then

$$n^{(q)}_{(2\nu+\delta)\times 2\nu_1}(S_{2\nu+\delta,\Delta}, S_{2\nu_1})$$
$$= q^{2\nu_1(\nu-\nu_1)+\delta\nu_1+\nu_1(\nu_1-1)} \prod_{i=\nu-\nu_1+1}^{\nu} (q^i - 1)(q^{i+\delta-1} + 1). \quad (85)$$

(ii) If

$$\nu \geqslant \begin{cases} \nu_1 + 1, & \text{when } \delta = 0, \text{ or } \delta = 1 \text{ and } \Delta_1 \neq \Delta, \\ \nu_1, & \text{when } \delta = 2, \text{ or } \delta = 1 \text{ and } \Delta_1 = \Delta, \end{cases} \quad (86)$$

then

$$n^{(q)}_{(2\nu+\delta)\times(2\nu_1+1)}(S_{2\nu+\delta,\Delta}, S_{2\nu_1+1,\Delta_1})$$
$$= n^{(q)}_{(2\nu+\delta)\times 2\nu_1}(S_{2\nu+\delta,\Delta}, S_{2\nu_1}) n^{(q)}_{(2(\nu-\nu_1)+\delta)\times 1}(S_{2(\nu-\nu_1)+\delta,\Delta}, \Delta_1), \quad (87)$$

where $n^{(q)}_{(2\nu+\delta)\times 2\nu_1}(S_{2\nu_1+\delta,\Delta}, S_{2\nu_1})$ is given by (85) and

$$n^{(q)}_{(2(\nu-\nu_1)+\delta)\times 1}(S_{2(\nu-\nu_1)+\delta,\Delta}, \Delta_1)$$
$$= \begin{cases} q^{\nu-\nu_1-1}(q^{\nu-\nu_1} - 1) & \text{when } \delta = 0, \\ q^{\nu-\nu_1}(q^{\nu-\nu_1} - 1) & \text{when } \delta = 1 \text{ and } \Delta_1 \neq \Delta, \\ q^{\nu-\nu_1}(q^{\nu-\nu_1} + 1) & \text{when } \delta = 1 \text{ and } \Delta_1 = \Delta, \\ q^{\nu-\nu_1}(q^{\nu-\nu_1+1} + 1) & \text{when } \delta = 2. \end{cases} \quad (88)$$

(iii) *If* $\nu \geqslant \nu_1 + (2 - \delta)$, *then*

$$n^{(q)}_{(2\nu+\delta)\times(2\nu_1+2)}(S_{2\nu+\delta,\Delta}, S_{2\nu_1+2,\Delta_1})$$
$$= n^{(q)}_{(2\nu+\delta)\times 2\nu_1}(S_{2\nu+\delta,\Delta}, S_{2\nu_1})\, n^{(q)}_{(2(\nu-\nu_1)+\delta)\times 2}\left(S_{2(\nu-\nu_1)+\delta,\Delta}, \begin{pmatrix} 1 & \\ & -z \end{pmatrix}\right), \tag{89}$$

where $n^{(q)}_{(2\nu+\delta)\times 2\nu_1}(S_{2\nu+\delta,\Delta}, S_{2\nu_1})$ *is given by* (85) *and*

$$n^{(q)}_{(2(\nu-\nu_1)+\delta)\times 2}\left(S_{2(\nu-\nu_1)+\delta,\Delta}, \begin{pmatrix} 1 & \\ & -z \end{pmatrix}\right)$$
$$= \begin{cases} q^{2(\nu-\nu_1)-2}(q^{\nu-\nu_1}-1)(q^{\nu-\nu_1}-1) & \text{when } \delta = 0, \\ q^{2(\nu-\nu_1)-1}(q^{\nu-\nu_1}-1)(q^{\nu-\nu_1}+1) & \text{when } \delta = 1, \\ q^{2(\nu-\nu_2)}(q^{\nu-\nu_1}+1)(q^{\nu-\nu_1+1}+1) & \text{when } \delta = 2. \end{cases} \tag{90}$$

We remark that formulas (85), (87), and (89) in Proposition 18 are actually the formulas in Lemmas 6.13, 6.16 and 6.19, respectively, of [5].

Proposition 19 *Let q be odd. Then*

$$n^{(q)}_{(2\nu+\delta)\times t}(S_{2\nu+\delta,\Delta}, 0^{(t)})$$
$$= \sum_{k=0}^{\min\{t,\nu\}} q^{(1/2)k(k-1)} \frac{\prod_{i=t-k+1}^{t}(q^i-1)\prod_{i=\nu-k+1}^{\nu}(q^i-1)(q^{i+\delta-1}+1)}{\prod_{i=1}^{k}(q^i-1)}. \tag{91}$$

We remark that in the proof of Proposition 19 the formula of the number of k-dimensional totally isotropic subspaces with respect to $S_{2\nu+\delta,\Delta}$ is used and this formula is due to Segre [4] (cf. also Corollary 6.23 of [5]).

Carlitz [2] obtained another formula for $n^{(q)}_{(2\nu+\delta)\times t}(S_{2\nu+\delta,\Delta}, 0^{(t)})$ in terms of certain terminating q-hypergeometric series, which can be stated as follows.

Proposition 20 *Let q be odd. Let $m = 2\nu + \delta$, where $\delta = 0, 1$, or 2. Then*

$$n^{(q)}_{m\times t}(S_{2\nu+\delta,\Delta}, 0^{(t)})$$

$$= \begin{cases} q^{mt-(1/2)t(t+1)}\{{}_2\Phi_0[q^{-(1/2)t}, q^{-(1/2)(t-1)}; q^{(1/2)m}]' \\ \quad -q^{-(1/2)m}(1-q^t){}_2\Phi_0[q^{-(1/2)(t-1)}, q^{-(1/2)(t-2)}; q^{(1/2)m}]'\} & \text{when } \delta = 0, \\ q^{mt-(1/2)t(t+1)}{}_2\Phi_0[q^{-(1/2)t}, q^{-(1/2)(t-1)}; q^{(1/2)(m+1)}]' & \text{when } \delta = 1, \\ q^{mt-(1/2)t(t+1)}\{{}_2\Phi_0[q^{-(1/2)t}, q^{-(1/2)(t-1)}; q^{(1/2)m}]' \\ \quad +q^{-(1/2)m}(1-q^t){}_2\Phi_0[q^{-(1/2)(t-1)}, q^{-(1/2)(t-2)}; q^{(1/2)m}]'\} & \text{when } \delta = 2. \end{cases}$$
(92)

6. Representations of Quadratic Forms (Even Characteristic Case)

In this section we assume that q is even. Let A and B be two $m \times m$ matrices over $\mathscr{F}_q$. Clearly, A and B represent the same quadratic form, i.e.,

$$(x_1, x_2, \cdots, x_m)\boldsymbol{A}^t(x_1, x_2, \cdots, x_m)$$
$$=(x_1, x_2, \cdots, x_m)\boldsymbol{B}^t(x_1, x_2, \cdots, x_m), \tag{93}$$

if and only if $A+B$ is alternate. Thus it is natural to introduce the congruence of $m \times m$ matrices modulo the set of $m \times m$ alternate matrices.

Denote the set of $m \times m$ alternate matrices over $\mathscr{F}_q$ by $\mathscr{H}_m$. Two $m \times m$ matrices A and B over $\mathscr{F}_q$ are said to be *congruent* mod $\mathscr{H}_m$ if $A+B \in \mathscr{H}_m$, which is usually denoted by

$$A \equiv B (\text{mod } \mathscr{H}_m), \tag{94}$$

or, simply,

$$A \equiv B. \tag{95}$$

The set of $m \times m$ matrices over $\mathscr{F}_q$ is partitioned into a disjoint union of congruent classes of matrices. Each congruent class of $m \times m$ matrices corresponds to a quadratic form in m variables in a one-to-one manner.

The $m \times m$ matrices A and B over $\mathscr{F}_q$ are said to be "*cogredient*" if there is a $P \in GL_m(\mathscr{F}_q)$ such that

$$^tPAP \equiv B. \tag{96}$$

Choose a fixed element $\alpha \in \mathscr{F}_q$ but $\alpha \notin N = \{x + x^2 | x \in \mathscr{F}_q\}$. Let

$$G_{2\nu} = \begin{pmatrix} 0 & I^{(\nu)} \\ 0 & \end{pmatrix}, \qquad (97)$$

$$G_{2\nu+1} = \begin{pmatrix} 0 & I^{(\nu)} \\ 0 & \\ & & 1 \end{pmatrix}, \qquad (98)$$

and

$$G_{2\nu+2} = \begin{pmatrix} 0 & I^{(\nu)} \\ 0 & \\ & & \alpha & 1 \\ & & & \alpha \end{pmatrix}. \qquad (99)$$

In order to cover these three cases we introduce the notation $G_{2\nu+\delta}$, where $\delta = 0, 1$, or 2. We call ν the *index* of $G_{2\nu+\delta}$. We recall that an $m \times m$ matrix A over $\mathscr{F}_q$ is "cogredient" to one of the normal forms

$$\begin{pmatrix} G_{2\nu+\delta} & \\ & 0^{(m-2\nu-\delta)} \end{pmatrix}, \quad \nu = 0, 1, 2, \cdots, \left[\frac{m-\delta}{2}\right], \qquad (100)$$

where $\delta = 0, 1$, or 2 (cf. Theorem 1.34 of [5]). We call ν the *index* of A and $2\nu + \delta$ the "*rank*" of A.

Let A be an $m \times m$ matrix over $\mathscr{F}_q$ and B be a $t \times t$ matrix over $\mathscr{F}_q$. Denote by $n_{m \times t}^{(q)}(A, B)$ the number of $m \times t$ matrices X such that

$$^tXAX \equiv B. \qquad (101)$$

Our problem now is to deduce formulas for $n_{m \times t}^{(q)}(A, B)$. As in Section 3 there is no loss of generality in assuming that one or both of A and B are in their normal forms under "cogredience" transformations.

Parallel to Propositions 6 and 16 we have

Proposition 21 *Let q be even. Let A be an $m \times m$ matrix over $\mathscr{F}_q$ and B be a $t \times t$ matrix over $\mathscr{F}_q$. Assume that A is of index ν and "rank" $2\nu + \delta$, where $\delta = 0, 1$, or 2 and $2\nu + \delta \leqslant m$. Then*

$$n_{m \times t}^{(q)}(A, B) = q^{t(m-2\nu-\delta)} n_{(2\nu+\delta) \times t}^{(q)}(G_{2\nu+\delta}, B). \qquad (102)$$

Thus our problem reduces to finding formulas for $n^{(q)}_{(2\nu+\delta)\times t}(G_{2\nu+\delta}, B)$.

Let P be a t-dimensional subspace of the $(2\nu+\delta)$-dimensional column space $\mathscr{F}_q^{(2\nu+\delta)}$. Any $(2\nu+\delta) \times t$ matrix of rank t, whose columns form a basis of P is called a *matrix representation of* P and is denoted also by P if no ambiguity arises.

We recall (cf. Section 7.1 of [5]) that a t-dimensional subspace P is said to be of *type* $(t, 2\nu_1+\delta_1, \nu_1)$ if ${}^t P G_{2\nu+\delta} P$ is cogredient to

$$\begin{pmatrix} G_{2\nu_1+\delta_1} & \\ & 0^{(t-2\nu_1-\delta_1)} \end{pmatrix}. \tag{103}$$

When $\delta = \delta_1 = 1$, subspaces of type $(t, 2\nu_1+1, \nu_1)$ are further subdivided into subspaces of type $(t, 2\nu_1+1, \nu_1, 0)$ and of type $(t, 2\nu_1+1, \nu_1, 1)$; a subspace P of type $(t, 2\nu_1+1, \nu_1)$ is said to be of *type*$(t, 2\nu_1+1, \nu_1, 0)$ or of *type*$(t, 2\nu_1+1, \nu_1, 1)$ if $e_{2\nu+1} \notin P$ or $e_{2\nu+1} \in P$, respectively. We introduce the notation

$$\Gamma_1 = \begin{cases} 0, & \text{if } \delta = \delta_1 = 1 \text{ and } e_{2\nu+1} \notin P, \\ 1, & \text{if } \delta = \delta_1 = 1 \text{ and } e_{2\nu+1} \in P, \\ \phi \text{ and may be omitted}, & \text{otherwise}. \end{cases} \tag{104}$$

Then the type of a t-dimensional subspace can be written as $(t, 2\nu_1+\delta_1, \nu_1, \Gamma_1)$. We know (cf. Theorem 7.5 of [5]) that subspaces of type $(t, 2\nu_1+\delta_1, \nu_1, \Gamma_1)$ exist in $\mathscr{F}_q^{(2\nu+\delta)}$ if and only if

$$2\nu_1 + \delta_1 \leqslant t$$
$$\leqslant \begin{cases} \nu + \nu_1 + \min\{\delta, \delta_1\} & \text{if } \delta \notin 1, \text{ or } \delta_1 \neq 1, \text{or } \delta = \delta_1 = 1 < \text{ and } \Gamma_1 = 1, \\ \nu + \nu_1 & \text{if } \delta = \delta_1 = 1 \text{ and } \Gamma_1 = 0. \end{cases}$$
$$\tag{105}$$

Define

$$O_{2\nu+\delta}(\mathscr{F}_q) = \{T \in GL_{2\nu+\delta}(\mathscr{F}_q) | {}^t T G_{2\nu+\delta} T \equiv G_{2\nu+\delta}\}, \tag{106}$$

which is called the *orthogonal group* of degree $2\nu+\delta$ over $\mathscr{F}_q$. We also know (cf. Theorem 7.6 of [5]) that two subspaces of $\mathscr{F}_q^{(2\nu+\delta)}$ belong to the same orbit of $O_{2\nu+\delta}(\mathscr{F}_q)$ if and only if they are of the same type. Subspaces of type $(2\nu_1 + \delta_1, 2\nu_1 + \delta_1, \nu_1, \Gamma_1)$ are called *nonsingular subspaces* and subspaces of type $(t, 0, 0)$ are called *totally singular subspaces*. For later reference we quote Theorem 7.3 and 7.10 of [5] as follows.

Lemma 22 *Let P be a nonsingular subspace of type $(2\nu_1 + \delta_1, 2\nu_1 + \delta_1, \nu_1)$, where $\delta_1 = 0$ or 2, in $\mathscr{F}_q^{(2\nu+d)}$. Then there exists a subspace P^* of $\mathscr{F}_q^{(2\nu+\delta)}$ such that $V_{2\nu+\delta} = P \cup P^*$, $P \cap P^* = (0)$ and ${}^tP(G + {}^tG)P^* = 0$. Furthermore, the "cogredience" normal form of ${}^tP^*GP^*$ is uniquely determined by G and the type of P (or the normal form of ${}^tPGP$) and is given in Table 1. In particular, the definite part of ${}^tP^*GP^*$ is of degree $|\delta - \delta_1|$.*

Lemma 23 *Let P_1 and P_2 be two $(2\nu + \delta) \times t$ matrices of rank t such that both P_1 and P_2 represent subspaces of type $(t, 2\nu_1 + \delta_1, \nu_1)$, where $\delta_1 = 0$ or 2. Then there exists an element $T \in O_{2\nu+\delta}(\mathscr{F}_q)$ such that $P_1 = TP_2$ if and only if ${}^tP_1 G_{2\nu+\delta} P_1 \equiv {}^tP_2 G_{2\nu+\delta} P_2$.*

Table 1

${}^tPGP$ \ G / ${}^tP^*GP^*$	$\begin{pmatrix} 0 & I^{(\nu)} \\ 0 & \end{pmatrix}$	$\begin{pmatrix} 0 & I^{(\nu)} \\ 0 & \\ & 1 \end{pmatrix}$	$\begin{pmatrix} 0 & I^{(\nu)} \\ 0 & \\ & \alpha & 1 \\ & & \alpha \end{pmatrix}$
$\begin{pmatrix} 0 & I^{(\nu_1)} \\ 0 & \end{pmatrix}$	$\begin{pmatrix} 0 & I^{(\nu-\nu_1)} \\ 0 & \end{pmatrix}$	$\begin{pmatrix} 0 & I^{(\nu-\nu_1)} \\ 0 & \\ & 1 \end{pmatrix}$	$\begin{pmatrix} 0 & I^{(\nu-\nu_1)} \\ 0 & \\ & \alpha & 1 \\ & & \alpha \end{pmatrix}$
$\begin{pmatrix} 0 & I^{(\nu_1)} \\ 0 & \\ & \alpha & 1 \\ & & \alpha \end{pmatrix}$	$\begin{pmatrix} 0 & I^{(\nu-\nu_1-2)} \\ 0 & \\ & \alpha & 1 \\ & & \alpha \end{pmatrix}$	$\begin{pmatrix} 0 & I^{(\nu-\nu_1-1)} \\ 0 & \\ & 1 \end{pmatrix}$	$\begin{pmatrix} 0 & I^{(\nu-\nu_1)} \\ 0 & \end{pmatrix}$

Parallel to Propositions 7 and 17, and by using Lemma 22, we deduce

Proposition 24 *Let q be even. Let B be a $t \times t$ matrix of index ν_1 and "rank" $2\nu_1 + \delta_1$.*

(i) *If the condition*
$$2\nu_1 + \delta_1 \leqslant t \leqslant \nu + \nu_1 + \min\{\delta, \delta_1\} \tag{107}$$
does not hold, then
$$n^{(q)}_{(2\nu+\delta)\times t}(G_{2\nu+\delta}, B) = 0. \tag{108}$$

(ii) *If $\delta_1 = 0$ and condition (107) holds, then*
$$n^{(q)}_{(2\nu+\delta)\times t}(G_{2\nu+\delta}, B)$$
$$= n^{(q)}_{(2\nu+\delta)\times 2\nu_1}(G_{2\nu+\delta}, G_{2\nu_1}) n^{(q)}_{(2(\nu-\nu_1)+\delta)\times(t-2\nu_1)}(G_{2(\nu-\nu_1)+\delta}, 0^{(t-2\nu_1)}). \tag{109}$$

(iii) *If $\delta_1 = 1$ and condition (107) holds, then*
$$n^{(q)}_{(2\nu+\delta)\times t}(G_{2\nu+\delta}, B)$$
$$= n^{(q)}_{(2\nu+\delta)\times 2\nu_1}(G_{2\nu+\delta}, G_{2\nu}) n^{(q)}_{(2(\nu-\nu_1)+\delta)\times(t-2\nu_1)}(G_{2(\nu-\nu_1)+\delta}, E_{11}), \tag{110}$$
where E_{11} is the $(t-2\nu_1)\times(t-2\nu_1)$ matrix with a 1 in its $(1,1)$ position and 0's elsewhere.

(iv) *If $\delta_1 = 2$ and condition (107) holds, then*
$$n^{(q)}_{(2\nu+\delta)\times t}(G_{2\nu+\delta}, B)$$
$$= n^{(q)}_{(2\nu+\delta)\times(2\nu_1+2)}(G_{2\nu-\delta}, G_{2\nu_1,2})$$
$$\times n^{(q)}_{(2(\nu-\nu_1)+\delta-2)\times(t-2\nu_1-2)}(G_{2(\nu-\nu_1-2+\delta)+2-\delta}, 0^{(t-2\nu_1-2)}). \tag{111}$$

Thus our problem reduces to computing of $n^{(q)}_{(2\nu+\delta)\times(2\nu_1+\delta_1)}(G_{2\nu+\delta}, G_{2\nu_1+\delta_1})$, $n^{(q)}_{(2\nu+\delta)\times t}(G_{2\nu+\delta}, 0^{(t)})$, and $n^{(q)}_{(2\nu+\delta)\times t}(G_{2\nu+\delta}, E_{11})$.

Parallel to Propositions 8 and 18 we have

Proposition 25 *Let q be even.*
(i) *If $\nu \geqslant \nu_1$, then*
$$n^{(q)}_{(2\nu+\delta)\times 2\nu_1}(G_{2\nu+\delta}, G_{2\nu_1})$$
$$= q^{2\nu_1(\nu-\nu_1)+\delta\nu_1+\nu_1(\nu_1-1)} \prod_{i=\nu-\nu_1+1}^{\nu}(q^i - 1)(q^{i+\delta-1} + 1). \tag{112}$$

(ii) *If*
$$\nu \geqslant \begin{cases} \nu_1 + 1, & \text{when } \delta = 0, \\ \nu_1, & \text{when } \delta = 1 \text{ or } 2, \end{cases} \tag{113}$$
then
$$n^{(q)}_{(2\nu+\delta)\times(2\nu_1+1)}(G_{2\nu+\delta}, G_{2\nu_1+1})$$
$$= n^{(q)}_{(2\nu+\delta)\times 2\nu_1}(G_{2\nu+\delta}, G_{2\nu_1}) n^{(q)}_{(2(\nu-\nu_1)+\delta)\times 1}(G_{2(\nu-\nu_1)+\delta}, G_1), \tag{114}$$
where $n^{(q)}_{(2\nu+\delta)\times 2\nu_1}(G_{2\nu+\delta}, G_{2\nu_1})$ *is given by* (112) *and*
$$n^{(q)}_{(2(\nu-\nu_1)+\delta)\times 1}(G_{2(\nu-\nu_1)+\delta}, G_1) = \begin{cases} q^{\nu-\nu_1-1}(q^{\nu-\nu_1}-1), & if\ \delta = 0, \\ q^{2(\nu-\nu_1)}, & if\ \delta = 1, \\ q^{\nu-\nu_1}(q^{\nu-\nu_1+1}+1), & if\ \delta = 2. \end{cases} \tag{115}$$

(iii) *If* $\nu \geqslant \nu_1 + (2 - \delta)$, *then*
$$n^{(q)}_{(2\nu+\delta)\times(2\nu_1+2)}(G_{2\nu+\delta}, G_{2\nu_1+2})$$
$$= n^{(q)}_{(2\nu+\delta)\times 2\nu_1}(G_{2\nu+\delta}, G_{2\nu_1}) n^{(q)}_{(2(\nu-\nu_1)+\delta)\times 2}(G_{2(\nu-\nu_1)+\delta}, G_2) \tag{116}$$
where $n^{(q)}_{(2\nu+\delta)\times 2\nu_1}(G_{2\nu+\delta}, G_{2\nu_1})$ *is given by* (112) *and*
$$n^{(q)}_{(2(\nu-\nu_1)+\delta)\times 2}(G_{2(\nu-\nu_1)+\delta}, G_2)$$
$$= \begin{cases} q^{2(\nu-\nu_1)-2}(q^{\nu-\nu_1-1}-1)(q^{\nu-\nu_1}-1), & if\ \delta = 0, \\ q^{2(\nu-\nu_1)-1}(q^{\nu-\nu_1}-1)(^{\nu-\nu_1}+1), & if\ \delta = 1, \\ q^{2(\nu-\nu_1)}(q^{\nu-\nu_1}+1)(q^{\nu-\nu_1+1}+1), & if\ \delta = 2. \end{cases} \tag{117}$$

We remark the formula (112) in the above proposition is formula (7.18) of [5] and the formulas (114) and (116) are special cases of formulas (7.24) and (7.27) of [5], respectively.

Parallel to Propositions 9 and 19 we derive in a similar way

Proposition 26 *Let q be even. Then*
$$n^{(q)}_{(2\nu+\delta)\times t}(G_{2\nu+\delta}, 0^{(t)})$$

$$= \sum_{k=0}^{\min\{t,\nu\}} q^{(1/2)k(k-1)} \frac{\prod_{i=t-k+1}^{t}(q^i-1) \prod_{i=\nu-k+1}^{\nu}(q^i-1)(q^{i+\delta-1}+1)}{\prod_{i=1}^{k}(q^i-1)}. \quad (118)$$

Corollary 27 *Let q be even. Let $m = 2\nu + \delta$, where $\delta = 0, 1$, or 2. Then*

$$n_{m\times t}^{(q)}(G_{2\nu+\delta}, 0^{(t)})$$
$$= \begin{cases} q^{mt-(1/2)t(t+1)}\{{}_2\Phi_0[q^{-(1/2)t}, q^{-(1/2)(t-1)}; q^{(1/2)m}]' \\ \quad - q^{-(1/2)m}(1-q^t){}_2\Phi_0[q^{-(1/2)(t-1)}, q^{-(1/2)(t-2)}; q^{(1/2)m}]'\}, & \text{when } \delta = 0, \\ q^{mt-(1/2)t(t+1)}{}_2\Phi_0[q^{-(1/2)t}, q^{-(1/2)(t-1)}; q^{(1/2)(m+1)}]', & \text{when } \delta = 1, \\ q^{mt-(1/2)t(t+1)}\{{}_2\Phi_0[q^{-(1/2)t}, q^{-(1/2)(t-1)}; q^{(1/2)m}]' \\ \quad + q^{-(1/2)m}(1-q^t){}_2\Phi_0[q^{-(1/2)(t-1)}, q^{-(1/2)(t-2)}; q^{(1/2)m}]'\}, & \text{when } \delta = 2. \end{cases}$$
$$(119)$$

Now let us come to the computation of $n_{(2\nu+\delta)\times t}^{(q)}(G_{2\nu+\delta}, E_{11})$.

Proposition 28 *Let q be even. Then*

$$n_{(2\nu+\delta)\times t}^{(q)}(G_{2\nu+\delta}, E_{11})$$
$$= \sum_{k=0}^{\min\{\nu,t-1\}} q^{(1/2)k(k+1)} \frac{\prod_{i=t-k}^{t-1}(q^i-1) \prod_{i=\nu-k+1}^{\nu}(q^i-1)(q^{i+\delta-1}+1)}{\prod_{i=1}^{k}(q^i-1)}$$
$$\times n_{(2(\nu-k)+\delta)\times 1}^{(q)}(G_{2(\nu-k)+\delta}, G_1), \quad (120)$$

where $n_{(2(\nu-k)+\delta)\times 1}^{(q)}(G_{2(\nu-k)+\delta}, G_1)$ is given by (115).

Proof It is the same thing to compute $n_{(2\nu+\delta)\times t}^{(q)}(G_{2\nu+\delta}, E_u)$. We know that $n_{(2\nu+\delta)\times t}^{(q)}(G_{2\nu+\delta}, E_u)$ is the number of $(2\nu+\delta) \times t$ matrix P such that

$$^tPG_{2\nu+\delta}P \equiv \begin{pmatrix} 0^{(t-1)} & \\ & 1 \end{pmatrix}. \quad (121)$$

Write
$$P = \begin{pmatrix} t-1 & 1 \\ P_1 & v; \end{pmatrix} \qquad (122)$$

then
$$^tP_1 G_{2\nu+\delta} P_1 \equiv 0, \qquad (123)$$

$$^tP_1(G_{2\nu+\delta} + {}^tG_{2\nu+\delta})v \equiv 0, \qquad (124)$$

and
$$^tv G_{2\nu+\delta} v = 1. \qquad (125)$$

Suppose that the columns of P_1 generate a subspace U of dimension k. Then U is totally singular and $k \leqslant \min\{\nu, t-1\}$. Let U be a matrix representation of the subspace U; then

$$P_1 = U T_{P_1}, \qquad (126)$$

where T_{P_1} is a $k \times (t-1)$ matrix of rank k and is uniquely determined by P_1 for a fixed choice of the matrix representation U. Since T_{P_1} is of rank k, (123) and (124) are equivalent to

$$^tU G_{2\nu+\delta} U \equiv 0 \qquad (127)$$

and
$$^tU(G_{2\nu+\delta} + {}^tG_{2\nu+\delta})v = 0, \qquad (128)$$

respectively. Therefore for any $(2\nu+\delta) \times (t-1)$ matrix P_1, whose columns generate a k-dimensional totally singular subspace U, the number of $(2\nu+\delta)$-dimensional column vectors v such that $(P_1 \ v)$ satisfies (123)–(125) depends only on U.

Now let U and V be two k-dimensional totally singular subspaces and P_1 and Q_1 be two $(2\nu+\delta) \times (t-1)$ matrices whose columns generate U and V, respectively. We assert that the number of $(2\nu+\delta)$-dimensional column vectors

v such that $(P_1\ v)$ satisfy (123)–(125). is equal to the number of column vectors u such that $(Q_1\ u)$ satisfy (123)–(125). Let

$$P_1 = UT_{P_1} \tag{129}$$

and

$$Q_1 = VT_{Q_1}, \tag{130}$$

where T_{P_1} and T_{Q_1} are $k \times (t-1)$ matrices of rank k. By the discussion of the above paragraph we can assume that $T_{P_1} = T_{Q_1}$. We have

$$^tUGU \equiv {}^tVGV \equiv 0; \tag{131}$$

then by Lemma 23 there is a $T \in O_{2\nu+\delta}(\mathscr{F}_q)$ such that

$$V = TU. \tag{132}$$

T_{P_1} is a $k \times (t-1)$ matrix of rank k, from which it follows that for a column vector v $(P_1\ v)$ satisfies (123)-(125) if and only if $(Q_1\ Tv)$ satisfies (123)–(125). This proves our assertion.

We know (cf. [4] or Corollary 7.25 of [5]) that the number of k-dimensional totally singular subspaces is equal to

$$\frac{\prod_{i=\nu-k+1}^{\nu}(q^i-1)(q^{i+\delta-1}+1)}{\prod_{i=1}^{k}(q^i-1)}, \tag{133}$$

and the number of $k \times (t-1)$ matrices of rank k can be obtained from (57) by replacing t by $t-1$. Thus it remains to compute the number of possible choices of column vectors v such that $(P_1\ v)$ satisfies (123)–(125), once P_1 is chosen. By the discussion of the above paragraph, we may choose

$$P_1 = \begin{pmatrix} I^{(k)} & \\ & 0^{(2\nu+\delta-k, t-1-k)} \end{pmatrix}, \tag{134}$$

and then compute the number of possible choices of v. From ${}^tP_1(G_{2\nu+\delta} + {}^tG_{2\nu+\delta})v = 0$ we deduce that v is of the form

$$v = {}^t(x_1, x_2, \cdots, x_\nu, \underbrace{0, 0, \cdots, 0}_{k}, x_{\nu+k+1}, \cdots, x_{2\nu+\delta}), \tag{135}$$

and the condition ${}^tvG_{2\nu+\delta}v=1$ becomes

$$\sum_{i=k+1}^{\nu} x_i x_{\nu+1} = 1, \quad \text{if } \delta = 0,$$

$$\sum_{i=k+1}^{\nu} x_i x_{\nu+i} + x_{2\nu+1}^2 = 1, \quad \text{if } \delta = 1, \tag{136}$$

$$\sum_{i=k+1}^{\nu} x_i x_{\nu+i} + \alpha x_{2\nu+1}^2 + x_{2\nu+1}x_{2\nu+2} + \alpha x_{2\nu+2}^2 = 1, \quad \text{if } \delta = 2.$$

Hence $x_1, x_2, \cdots, x_k$ can be chosen arbitrarily in $\mathscr{F}_q$ and the number of possible choices of $x_{k+1}, \cdots, x_\nu, x_{\nu+k+1}, \cdots, x_{2\nu+\delta}$ is $n_{(2(\nu-k)+\delta) \times 1}^{(q)}(G_{2(\nu-k)+\delta}, G_1)$, which is given by (115). Therefore once P_1 is chosen, the number of possible choices of v is

$$q^k n_{(2(\nu-k)+\delta) \times 1}^{(q)}(G_{2(\nu-k)+\delta}, G_1). \tag{137}$$

Proposition 28 is now proved. $\square$

7. Representations of Symmetric Bilinear Forms

In this section we assume again that q is even. Let A be an $m \times m$ symmetric matrix over $\mathscr{F}_q$ and B be a $t \times t$ symmetric matrix over $\mathscr{F}_q$. Denote by $n_{m \times t}^{(s)}(A, B)$ the number of $m \times t$ matrices X such that

$$ {}^tXAX = B. \tag{138}$$

Our problem is to deduce formulas for $n_{m \times t}^{(s)}(A, B)$. Since q is even, symmetric matrices are either alternate or nonalternate. If A is alternate and B is nonalternate, then, clearly, $n_{m \times t}^{(s)}(A, B) = 0$. If both A and B are alternate, then this is the case we studied in Section 3. Therefore we need to consider only the case when A is nonalternate.

As in Section 3 there is no loss of generality in assuming that one or both of A and B are in their normal forms under cogredience transformations. Let

$$S_{2\nu+1} = \begin{pmatrix} 0 & I^{(\nu)} & \\ I^{(\nu)} & 0 & \\ & & 1 \end{pmatrix} \tag{139}$$

and

$$S_{2\nu+2} = \begin{pmatrix} 0 & I^{(\nu)} & & \\ I^{(\nu)} & 0 & & \\ & & 0 & 1 \\ & & 1 & 1 \end{pmatrix}. \tag{140}$$

It is convenient to introduce the notation $S_{2\nu+\delta}$, where $\delta = 1$ or 2. We recall that an $m \times m$ nonalternate symmetric matrix A is cogredient to one of the normal forms

$$\begin{pmatrix} S_{2\nu+\delta} & \\ & 0^{(m\ 2\nu-\delta)} \end{pmatrix}, \tag{141}$$

where $\delta = 1$ or 2 and $\nu = 0, 1, 2, \cdots, [(m-\delta)/2]$ (cf. Theorem 4.1 of [5]). We call ν the *index* of A.

Parallel to Propositions 6 and 7 of Section 3 we have

Proposition 29 Let q be even. Let A be an $m \times m$ nonalternate symmetric matrix of index ν and rank $2\nu + \delta$, where $\delta = 1$ or 2 and $2\nu + \delta \leqslant m$, and let B be a $t \times t$ symmetric matrix, both over $\mathscr{F}_q$. Then

$$n^{(s)}_{m \times t}(A, B) = q^{t(m-2\nu-\delta)} n^{(s)}_{(2\nu+\delta) \times t}(S_{2\nu+\delta}, B). \tag{142}$$

Proposition 30 Let q be even. Let $\nu \geqslant 0$, $\delta = 1$ or 2, and B be a $t \times t$ symmetric matrix over $\mathscr{F}_q$. Assume that B is of index ν_1 and rank $2\nu_1 + \delta_1$, where $\delta_1 = 0, 1$, or 2 and $2\nu_1 + \delta_1 \leqslant t$.

(i) If $\nu_1 > \nu$, or $\nu_1 = \nu$ and $\delta_1 > \delta$, then

$$n^{(q)}_{(2\nu+\delta) \times t}(S_{2\nu+\delta}, B) = 0. \tag{143}$$

(ii) If $\nu_1 = \nu$ and $\delta_1 \leqslant \delta$, or $\nu_1 < \nu$, then there exist ν_2 and δ_2, where $\delta_2 = 0, 1$, or 2, such that

$$\begin{pmatrix} S_{2\nu_1+\delta_1} & \\ & S_{2\nu_2+\delta_2} \end{pmatrix} \tag{144}$$

is cogredient to $S_{2\nu+\delta}$ and

$$n^{(s)}_{(2\nu+\delta)\times t}(S_{2\nu+\delta}, B)$$
$$= n^{(s)}_{(2\nu+\delta)\times(2\nu_1+\delta_1)}(S_{2\nu+\delta}, S_{2\nu_1+\delta_1}) n^{(s)}_{(2\nu_2+\delta_2)\times(t-2\nu_1-\delta_1)}(S_{2\nu_2+\delta_2}, 0^{(t-2\nu_1-\delta_1)}). \tag{145}$$

Let $\nu \geqslant 0$ and $\delta = 1$ or 2. Define

$$PS_{2\nu+\delta}(\mathscr{F}_q) = \{T \in GL_{2\nu+\delta}(\mathscr{F}_q) | {}^t T S_{2\nu+\delta} T = S_{2\nu+\delta}\}, \tag{146}$$

which is called the *pseudo-symplectic group* of degree $2\nu + \delta$ over $\mathscr{F}_q$ (cf. Section 4.1 of [5]). By Theorem 4.7 of [5] we have

$$|PS_{2\nu+\delta}(\mathscr{F}_q)| = q^{(\nu+\delta-1)^2} \prod_{i=1}^{\nu} (q^{2i} - 1). \tag{147}$$

We recall (cf. Section 4.2 of [5]) that a t-dimensional subspace P of the $(2\nu+\delta)$-dimensional column vector space $\mathscr{F}_q^{(2\nu+\delta)}$ is of $type(t, 2\nu_1 + \delta_1, \nu_1, \varepsilon_1)$, where $\delta_1 = 0, 1$, or 2, $2\nu_1 + \delta_1 \leqslant t$, and $\varepsilon_1 = 0$ or 1, if ${}^t P S_{2\nu+\delta} P$ is cogredient to

$$\begin{pmatrix} S_{2\nu_1+\delta_1} & \\ & 0^{(t-2\nu_1-\delta_1)} \end{pmatrix}, \tag{148}$$

and $\varepsilon_1 = 0$ or 1 according to $e_{2\nu+1} \notin P$ or $e_{2\nu+1} \in P$, respectively. In particular, subspaces of type $(2\nu_1+\delta_1, 2\nu_1+\delta_1, \nu_1, \varepsilon_1)$ are called *nonisotropic subspaces* and subspaces of type $(t, 0, 0, \varepsilon_1)$ are called *totally isotropic subspaces*. It is known that subspaces of type $(t, 2\nu_1 + \delta_1, \nu_1, \varepsilon_1)$ exist if and only if

$$(\delta_1, \varepsilon_1) = \begin{cases} (0,0), (1,0), (1,1), \text{ or } (2,0), & \text{if } \delta = 1, \\ (0,0), (0,1), (1,0), (2,0) \text{ or } (2,1), & \text{if } \delta = 2, \end{cases} \tag{149}$$

and
$$2\nu_1 + \max\{\delta_1, \varepsilon_1\} \leqslant t \leqslant \nu + \nu_1 + [(\delta + \delta_1 - 1)/2] + \varepsilon_1, \tag{150}$$

and that subspaces of the same type constitute an orbit of subspaces under $PS_{2\nu+\delta}(\mathscr{F}_q)$ (cf. Theorems 4.11 and 4.12 of [5]). Denote the number of subspaces of type $(t, 2\nu_1+\delta_1, \nu_1, \varepsilon_1)$ by $N(t, 2\nu_1+\delta_1, \nu_1, \varepsilon_1; 2\nu+\delta)$; then a closed formula of $N(t, 2\nu_1+\delta_1, \nu_1, \varepsilon_1; 2\nu+\delta)$ is given in Theorem 4.14 of [5] and reads as

$$\begin{aligned}&N(t, 2\nu_1 + \delta_1, \nu_1, \varepsilon_1; 2\nu + \delta)\\&= q^{n_0 + 2(\nu_1 + (2-\delta)[\delta_1/2])(\nu+\nu_1-t+\delta[(\delta_1+1)/2]+(\delta-1)(\delta_1-1)(\delta_1-2)\varepsilon_1/2)}\\&\quad \times \frac{\prod\limits_{i=\nu+\nu_1-t+[(\delta+\delta_1-1)/2]+\varepsilon_1+1}^{\nu}(q^{2i}-1)}{\prod\limits_{i=1}^{\nu_1}(q^{2i}-1) \prod\limits_{i=1}^{t-2\nu_1-\max\{\delta_1,\varepsilon_1\}}(q^i-1)},\end{aligned} \tag{151}$$

where $n_0 = 0$ if $\delta = 1$, and $n_0 = t, 0, 2(\nu+1)-t, 2(\nu+1)-t$, or $2(\nu+1)-t$ corresponding to the cases $(\delta_1, \varepsilon_1) = (0,0), (0,1), (1,0), (2,0)$, or $(2,1)$, respectively, if $\delta=2$.

Proposition 31 *Let q be even. Let $\delta = 1$ or 2 and $\delta_1 = 0, 1$, or 2. Assume that $\nu = \nu_1$ and $\delta_1 \leqslant \delta$, or $\nu_1 < \nu$. Then*

$$\begin{aligned}&n^{(s)}_{(2\nu+\delta)\times(2\nu_1+\delta_1)}(S_{2\nu+\delta}, S_{2\nu_1+\delta_1})\\&= |PS_{2\nu_1+\delta_1}(\mathscr{F}_q)| \sum_{\varepsilon_1=0}^{1} N(2\nu_1+\delta_1, 2\nu_1+\delta_1, \nu_1, \varepsilon_1; 2\nu+\delta),\end{aligned} \tag{152}$$

where $|PS_{2\nu_1+\delta_1}(\mathscr{F}_q)|$ is given by (147) and $N(2\nu_1+\delta_1, 2\nu_1+\delta_1, \nu_1, \varepsilon_1; 2\nu+\delta)$ is given by (151).

Proof Let X be a $(2\nu+\delta) \times (2\nu_1+\delta_1)$ matrix over $\mathscr{F}_q$ satisfying

$$^tX S_{2\nu+\delta} X = S_{2\nu_1+\delta_1}. \tag{153}$$

Then rank $X = 2\nu_1 + \delta_1$ and the column vectors of X span a $(2\nu_1+\delta_1)$-dimensional subspace of $\mathscr{F}_q^{(2\nu+\delta)}$. This subspace is either of type $(2\nu_1+\delta_1, 2\nu_1+$

$\delta_1, \nu_1, 0$) or of type $(2\nu_1 + \delta_1, 2\nu_1 + \delta_1, \nu_1, 1)$. Conversely, let U be a subspace of type $(2\nu_1 + \delta_1, 2\nu_1 + \delta_1, \nu_1, \varepsilon_1)$, where $\varepsilon_1 = 0$ or 1; then we can find a $(2\nu + \delta) \times (2\nu_1 + \delta_1)$ matrix X such that the column vectors form a basis of U and (153) holds. Let X and X_1 be two $(2\nu + \delta) \times (2\nu_1 + \delta_1)$ matrices such that both of them satisfy (153); then the columns of X and the columns of X_1 span the same subspace if and only if there is a $T \in PS_{2\nu_1+\delta_1}(\mathscr{F}_q)$ such that $TX = X_1$. Hence (152) follows immediately. □

Proposition 32 *Let q be even. Let $\delta = 1$ or 2. Then*

$$n^{(s)}_{(2\nu+\delta) \times t}(S_{2\nu+\delta}, 0^{(t)})$$
$$= \begin{cases} \sum_{k=0}^{\min\{t,\nu\}} q^{(1/2)k(k-1)} \prod_{i=t-k+1}^{t} (q^i - 1) N(k, 0, 0, 0; 2\nu + 1), & \text{if } \delta = 1, \\ \sum_{k=0}^{\min\{t,\nu+1\}} q^{(1/2)k(k-1)} \prod_{i=t-k+1}^{t} (q^i - 1) \sum_{\varepsilon_1=0}^{l} N(k, 0, 0, \varepsilon_1; 2\nu + 2), & \text{if } \delta = 2, \end{cases}$$
(154)

where $N(k, 0, 0, 0; 2\nu + 1)$ and $N(K, 0, 0, \varepsilon_1; 2\nu + 2)$ are given by (151).

References

[1] L. Carlitz, Representations by quadratic forms in a finite field, *Duke Math.J.* 21 (1954), 123-137.

[2] L. Carlitz, Representations by skew forms in a finite field, *Arch. Math.* 5 (1954), 19-31.

[3] L. Carlitz and J. H. Hodges, Representations by hermitian forms in a finite field, *Duke Math. J.* 22 (1955), 393-405.

[4] B. Segre, On Galois geometries, *in* "Proceedings of International Congress of Mathematicians, Edinburgh, 1958," pp. 488-499. Cambridge University Press, Cambridge, UK, 1960.

[5] Z. Wan. "Geometry of Classical Groups over Finite Fields," Studentlitteratur, Lund. 1993.

Geometry of Matrices

Zhexian Wan

In Memory of Professor L. K. Hua (1910–1985)

1. Introduction

The study of the geometry of matrices was initiated by L. K. Hua in the mid forties [5–10]. At first, relating to his study of the theory of functions of several complex variables, he began studying four types of geometry of matrices over the complex field, i.e., geometries of rectangular matrices, symmetric matrices, skew-symmetric matrices, and hermitian matrices. In 1949, he [11] extended his result on the geometry of symmetric matrices over the complex field to any field of characteristic not 2, and in 1951 he [12] extended his result on the geometry of rectangular matrices to any division ring distinct from $\mathbb{F}_2$ and applied it to problems in algebra and geometry. Then the study of the geometry of matrices was succeeded by many mathematicians. In recent years it has also been applied to graph theory.

To explain the problems of the geometry of matrices we are interested in, it is better to start with the Erlangen Program which was formulated by F. Klein in 1872. It says: "A geometry is the set of properties of figures which are invariant under the nonsingular linear transformations of some group". There F. Klein pointed out the intimate relationship between geometry, group, and

---

Received February 28, 1995.

Revised May 16, 1995.

This paper was presented at the International Conference on Algebraic Combinatorics, Fukuoka, Japan, November 22–26, 1993.

invariants. Then a fundamental problem in a geometry in the sense of Erlangen Program is to characterize the transformation group of the geometry by as few geometric invariants as possible. The answer to this problem is often called the fundamental theorem of the geometry.

In a geometry of matrices, the points of the associated space are a certain kind of matrices of the same size, and there is a transformation group acting on this space. Take the geometry of rectangular matrices as an example. Let D be a division ring, and m and n be integers $\geqslant 2$. The *space* of the geometry of rectangular matrices over D consists of all $m \times n$ matrices over D and is denoted by $\mathcal{M}_{m \times n}(D)$. The elements of $\mathcal{M}_{m \times n}(D)$ are called the *points* of the space. $\mathcal{M}_{m \times n}(D)$ admits transformations of the following form

$$\mathcal{M}_{m \times n}(D) \to \mathcal{M}_{m \times n}(D)$$
$$X \mapsto PXQ + R, \tag{1}$$

where $P \in GL_m(D)$, $Q \in GL_n(D)$, and $R \in \mathcal{M}_{m \times n}(D)$. All these transformations form a *transformation group* of $\mathcal{M}_{m \times n}(D)$, which is denoted by $G_{m \times n}(D)$. Then the geometry of rectangular matrices aims at the study of the invariants of its geometric figures (or subsets) under $G_{m \times n}(D)$. For instance, for the figure formed by two $m \times n$ matrices X_1 and X_2 over D, rank $(X_1 - X_2)$ is an invariant under $G_{m \times n}(D)$. If rank $(X_1 - X_2) = 1$, X_1 and X_2 are called *adjacent*. L. K. Hua proved that the invariant "adjacency" alone is "almost" sufficient to characterize the transformation group $G_{m \times n}(D)$ of $\mathcal{M}_{m \times n}(D)$, which will be explained in detail in the next section.

2. Geometry of Rectangular Matrices

Fundamental Theorem of the Geometry of Rectangular Matrices. Let D be a division ring, m and n integers $\geqslant 2$, $\mathcal{A}$ a bijective map from $\mathcal{M}_{m \times n}(D)$ to itself. Assume that both $\mathcal{A}$ and $\mathcal{A}^{-1}$ preserve the adjacency, i.e., for any two points X_1 and X_2 of $M_{m \times n}(D)$, X_1 and X_2 are adjacent if and only if $\mathcal{A}(X_1)$ and $\mathcal{A}(X_2)$ are adjacent. Then, when $m \neq n$, $\mathcal{A}$ is of the form

$$\mathcal{A}(X) = PX^\sigma Q + R \text{ for all } X \in \mathcal{M}_{m \times n}(D), \tag{2}$$

where $P \in GL_m(D)$, $Q \in GL_n(D)$, $R \in \mathcal{M}_{m \times n}(D)$, σ is an automorphism of D, and X^σ is the matrix obtained from X by applying σ to all its entries. When $m = n$, besides (1) $\mathcal{A}$ can also be of the form

$$\mathcal{A}(X) = P^t(X^\tau)Q + R \text{ for all } X \in \mathcal{M}_{m \times n}(D), \tag{3}$$

where P, Q, and R have the same meaning as above, and τ is an antiautomorphism of D. Conversely, both maps (2) and (3) are bijections, and they and their inverses preserve the adjacency. □

When $D \neq \mathbb{F}_2$, the theorem was proved by L. K. Hua [12] in 1951. The proof for the case $D = \mathbb{F}_2$ was supplemented by Z. Wan and Y. Wang [24] in 1962. The key tool to prove this theorem is the maximal set introduced by L. K. Hua. A *maximal set* in $\mathcal{M}_{m \times n}(D)$ is a maximal set of points such that any two of them are adjacent. Thus the concept of a maximal set is actually the concept of a maximal clique appeared in graph theory twenty years later. Clearly a bijective map $\mathcal{A}$ for which both $\mathcal{A}$ and $\mathcal{A}^{-1}$ preserve the adjacency carries maximal sets into maximal sets. The main steps Hua used to prove the above theorem is as follows: First he determined the normal forms of maximal sets under $G_{m \times n}(D)$. They are

$$\left\{ \begin{pmatrix} x_{11} & x_{12} & \cdots & x_{1n} \\ 0 & 0 & \cdots & 0 \\ \vdots & \vdots & & \vdots \\ 0 & 0 & \cdots & 0 \end{pmatrix} \middle| x_{11}, x_{12}, \cdots, x_{1n} \in D \right\} \tag{4}$$

and

$$\left\{ \begin{pmatrix} x_{11} & 0 & \cdots & 0 \\ x_{21} & 0 & \cdots & 0 \\ \vdots & \vdots & & \vdots \\ x_{m1} & 0 & \cdots & 0 \end{pmatrix} \middle| x_{11}, x_{21}, \cdots, x_{m1} \in D \right\}. \tag{5}$$

Then by defining the intersection of two maximal sets which contain two adjacent points in common to be a *line* in any one of them, he proved that $\mathcal{A}$ induces bijective maps on maximal sets, which carries lines into lines and that a line in the maximal set (4) is of the form

$$\left\{ \begin{pmatrix} ta_{11}+b_{11} & ta_{12}+b_{12} & \cdots & ta_{1n}+b_{1n} \\ 0 & 0 & \cdots & 0 \\ \vdots & \vdots & & \vdots \\ 0 & 0 & \cdots & 0 \end{pmatrix} \Bigg| \, t \in D \right\}, \qquad (6)$$

where $a_{11}, a_{12}, \cdots, a_{1n}, b_{11}, b_{12}, \cdots, b_{1n} \in D$. When $D \neq \mathbb{F}_2$, by the fundamental theorem of affine geometry, after subjecting $\mathcal{A}$ to a bijective map of the form (2) or (3) (which will be needed only when $m = n$), it can be assumed that $\mathcal{A}$ leaves both the maximal sets (4) and (5) pointwise fixed. Finally it can be proved that $\mathcal{A}$ leaves every point of $\mathcal{M}_{m \times n}(D)$ fixed.

In [12], from the above theorem L. K. Hua deduced the explicit forms of automorphisms, semi-automorphisms, Jordan automorphisms, and Lie automorphisms of the total matrix ring $\mathcal{M}_n(D)(n \geqslant 2)$ over D. For Jordan automorphisms it is assumed that the characteristic of D is not 2, and for Lie automorphisms it is assumed that the characteristic of D is not 2 and 3. He also deduced the fundamental theorem of the projective geometry of rectangular matrices over D (for detailed proof, cf. [17]). When D is a field, the latter was proved by W. L. Chow [2] in 1949. In 1965, S. Deng and Q. Li [3] deduced the fundamental theorem of the geometry of rectangular matrices over a field from Chow's result.

Call the points of $\mathcal{M}_{m \times n}(D)$ the *vertices* and define two vertices *adjacent* if they are adjacent points. Then a *graph* is obtained. Denote this graph by $\Gamma(\mathcal{M}_{m \times n}(D))$. Naturally, the fundamental theorem of the geometry of rectangular matrices can be interpreted as a theorem on graph automorphisms of $\Gamma(\mathcal{M}_{m \times n}(D))$ [1].

3. Geometry of Alternate Matrices

In this section we assume that F is a field and n is an integer $\geqslant 2$. Let A be an $n \times n$ matrix over F. If $^tA = -A$ and all entries along the main diagonal of A are 0's, then A is called an $n \times n$ *alternate matrix* over F. Denote by $\mathcal{K}_n(F)$ the set of all $n \times n$ alternate matrices over F, and call it the *space* of the geometry of $n \times n$ alternate matrices and its elements the *points*. Transformations of $\mathcal{K}_n(F)$ to itself of the following form

$$\mathcal{K}_n(F) \to \mathcal{K}_n(F)$$
$$X \mapsto {^tPXP} + K, \tag{7}$$

where $P \in GL_n(F)$ and $K \in \mathcal{K}_n(F)$, form a *transformation group* of $\mathcal{K}_n(F)$, denoted by $G\mathcal{K}_n(F)$. Let X_1 and $X_2 \in \mathcal{K}_n(F)$. If rank $(X_1 - X_2) = 2$, then X_1 and X_2 are said to be *adjacent*. Clearly, the adjacency is an invariant under $G\mathcal{K}_n(F)$. Conversely, we have

Fundamental Theorem of the Geometry of Alternate Matrices.
Let F be a field of any characteristic, n an integer $\geqslant 4$, and $\mathcal{A}$ a bijective map from $\mathcal{K}_n(F)$ to itself. Assume that both $\mathcal{A}$ and $\mathcal{A}^{-1}$ preserve the adjacency. Then, when $n > 4$, $\mathcal{A}$ is of the form

$$\mathcal{A}(X) = a\,{^tP}X^\sigma P + K \text{ for all } X \in \mathcal{K}_n(F), \tag{8}$$

where $a \in F^*$, $P \in GL_n(F)$, $K \in \mathcal{K}_n(F)$, and σ is an automorphism of F. When $n = 4$, $\mathcal{A}$ is of the form

$$\mathcal{A}(X) = a\,{^tP}(X^*)^\sigma P + K \text{ for all } X \in \mathcal{K}_4(F), \tag{9}$$

where a, P, K, and σ have the same meaning as above, and $X \to X^*$ is either the identity map of $\mathcal{K}_4(F)$ or the following map

$$\begin{pmatrix} 0 & x_{12} & x_{13} & x_{14} \\ -x_{12} & 0 & x_{23} & x_{24} \\ -x_{13} & -x_{23} & 0 & x_{34} \\ -x_{14} & -x_{24} & -x_{34} & 0 \end{pmatrix} \mapsto \begin{pmatrix} 0 & x_{12} & x_{13} & x_{23} \\ -x_{12} & 0 & x_{14} & x_{24} \\ -x_{13} & -x_{14} & 0 & x_{34} \\ -x_{23} & -x_{24} & -x_{34} & 0 \end{pmatrix}. \tag{10}$$

Conversely, both maps (8) and (9) are bijective, and they and their inverses preserve the adjacency. Q.E.D.

The above theorem was proved by M. Liu [16] in 1966, the proof relies also on the concept of *maximal sets*. When $F = \mathbb{C}$ and $\mathcal{A}$ satisfies further conditions, it was proved by L. K. Hua [5] in 1945. The map (10) was also discovered by L. K. Hua [5] in 1945.

This theorem has also applications to algebra and geometry [16], and can also be interpreted as a theorem on graph automorphisms [1].

4. Geometry of Symmetric Matrices

In this section we assume again that F is a field and n is an integer $\geqslant 2$. An $n \times n$ matrix S over F is called *symmetric* if ${}^t S = S$. Denote by $\mathcal{S}_n(F)$ the set of all $n \times n$ symmetric matrices over F, and call it the *space* of the geometry of $n \times n$ symmetric matrices and its elements the *points*. The set of all transformations of $\mathcal{S}_n(F)$ to itself of the form

$$\mathcal{S}_n(F) \to \mathcal{S}_n(F)$$
$$X \mapsto {}^t PXP + S, \qquad (11)$$

where $P \in GL_n(F)$ and $S \in \mathcal{S}_n(F)$, forms a *transformation group* of $\mathcal{S}_n(F)$, denoted by $GS_n(F)$. Let $X_1, X_2 \in \mathcal{S}_n(F)$. When rank $(X_1 - X_2) = 1$, then X_1 and X_2 are said to be *adjacent*. Clearly, the adjacency of two points in $\mathcal{S}_n(F)$ is an invariant under $GS_n(F)$. Conversely, we have

Fundamental Theorem of the Geometry of Symmetric Matrices. Let F be a field of any characteristic and n be an integer $\geqslant 2$; when F is of characteristic two and $F \neq \mathbb{F}_2$ we assume further that $n \geqslant 3$. Let $\mathcal{A}$ be a bijective map from $\mathcal{S}_n(F)$ to itself and assume that both $\mathcal{A}$ and $\mathcal{A}^{-1}$ preserve the adjacency. Then unless $n = 3$ and $F = \mathbb{F}_2$, $\mathcal{A}$ is of the form

$$\mathcal{A}(X) = a {}^t P X^\sigma P + S \text{ for all } X \in \mathcal{S}_n(F), \qquad (12)$$

where $a \in F^*$, $P \in GL_n(F)$, $S \in \mathcal{S}_n(F)$, and σ is an automorphism of F. When $n = 3$ and $F = \mathbb{F}_2$, the bijective map

$$\begin{cases} \begin{pmatrix} x_{11} & x_{12} & x_{13} \\ x_{12} & x_{22} & 0 \\ x_{13} & 0 & x_{33} \end{pmatrix} \mapsto \begin{pmatrix} x_{11} & x_{12} & x_{13} \\ x_{12} & x_{22} & 0 \\ x_{13} & 0 & x_{33} \end{pmatrix} \\ \begin{pmatrix} x_{11} & x_{12} & x_{13} \\ x_{12} & x_{22} & 1 \\ x_{13} & 1 & x_{33} \end{pmatrix} \mapsto \begin{pmatrix} x_{11}+1 & x_{12}+1 & x_{13}+1 \\ x_{12}+1 & x_{22} & 1 \\ x_{13}+1 & 1 & x_{33} \end{pmatrix} \end{cases} \quad (13)$$

from $\mathcal{S}_3(\mathbb{F}_2)$ to itself preserves also the adjacency and $\mathcal{A}$ is a product of maps of the form (12) or (13). Q.E.D.

When $F = \mathbb{C}$ and $\mathcal{A}$ satisfies further conditions, the above theorem was first proved by L. K. Hua [5] in 1945. In 1949 he [11] proved the theorem for any field of characteristic not two by the method of constructing involutions. But there are some gaps in his paper [11] which the author could not fill in. Without any restriction on the characteristic of F the author [18, 19] proved the above theorem. In the proof, besides the maximal sets which were defined in the same way as in the geometry of rectangular matrices and were called the *maximal sets of rank* 1 by the author, the *maximal sets of rank* 2 were also introduced. At first, the *distance* $d(X,Y)$ between two points X and Y of $\mathcal{S}_n(F)$ is defined to be the least integer d such that there is a sequence of $d+1$ points

$$X_0 = X, X_1, X_2, \cdots, X_d = Y$$

of $\mathcal{S}_n(F)$ for which any pair of consecutive points X_i and X_{i+1} ($i = 0, 1, 2, \cdots, d-1$) are adjacent. Assume that F is of characteristic not two. Then a subset $\mathcal{L}$ of $\mathcal{S}_n(F)$ is called a *maximal set of rank* 2 if (i) $\mathcal{L}$ contains a maximal set of rank 1, denoted by $\mathcal{M}$, (ii) of any $S \in \mathcal{L} \setminus \mathcal{M}$ and $M \in \mathcal{M}$, $d(S, M) = 2$, and (iii) for any $T \in \mathcal{S}_n(F)$, $d(T, M) = 2$ for all $M \in \mathcal{M}$ implies $T \in \mathcal{L}$. When F is characteristic two, the definition of maximal sets of rank 2 should be modified

[19]. Clearly, if $\mathcal{A}$ is a bijective map of $\mathcal{S}_n(F)$ for which both $\mathcal{A}$ and $\mathcal{A}^{-1}$ preserve the adjacency, then $\mathcal{A}$ carries maximal sets of rank 1 into maximal sets of rank 1 and maximal sets of rank 2 into maximal sets of rank 2. The normal form of maximal sets of rank 1 under $GS_n(F)$ is

$$\left\{ \left(\begin{array}{cccc} x & 0 & \cdots & 0 \\ 0 & 0 & \cdots & 0 \\ \vdots & \vdots & & \vdots \\ 0 & 0 & \cdots & 0 \end{array} \right\vert x \in F \right) \right\}, \tag{14}$$

and the normal form of maximal sets of rank 2 under $GS_n(F)$ is

$$\left\{ \left(\begin{array}{cccc} x_{11} & x_{12} & \cdots & x_{1n} \\ x_{12} & 0 & \cdots & 0 \\ \vdots & \vdots & & \vdots \\ x_{1n} & 0 & \cdots & 0 \end{array} \right\vert x_{11}, x_{12}, \cdots, x_{1n} \in F \right) \right\}. \tag{15}$$

Then maximal sets of rank 2 are used in the proof of the above theorem instead of the maximal sets used in the proof of the fundamental theorem of the geometry of rectangular matrices. The case when $n = 2$, F is of characteristic two, and $F \neq \mathbb{F}_2$ still remains open.

When F is of characteristic not two, from the above theorem we can deduce the explicit form of the automorphisms of the Jordan ring of $n \times n$ symmetric matrices over F [18] and the fundamental theorem of the dual polar space of type C_n due to W. L. Chow [2] (cf. [15], [23]).

Call the points of $\mathcal{S}_n(F)$ *vertices*. Two vertices are said to be *adjacent* if they are adjacent as points. Then we obtain a *graph*, denoted by $\Gamma(\mathcal{S}_n(F))$. The fundamental theorem of the geometry of symmetric matrices can naturally be interpreted as a theorem on graph automorphisms of the graph $\Gamma(\mathcal{S}_n(F))$ [18, 19].

It is interesting that when F is a finite field of characteristic not two and $n \geqslant 2$, and when F is a finite field of characteristic two and $n \geqslant 3$, besides $\Gamma(\mathcal{S}_3(\mathbb{F}_2))$, all $\Gamma(\mathcal{S}_n(F))$ are not distance-transitive. But the author [20] proved that $\Gamma(\mathcal{S}_3(\mathbb{F}_2))$ is distance-transitive, hence, distanceregular, and isomorphic to the graph of the folded 7-cube.

Now assume that F is of characteristic not two. Let $X_1, X_2 \in \mathcal{S}_n(F)$. When rank $(X_1 - X_2) = 1$ or 2, we say that X_1 and X_2 are adjacent. Then we obtain also a *graph*, denoted by $\Gamma^*(\mathcal{S}_n(F))$. From the fundamental theorem of the geometry of symmetric matrices we can deduce that the graph automorphisms of $\Gamma^*(\mathcal{S}_n(F))$ are of the form (12) (cf. [18]). When $F \neq \mathbb{F}_q$, the graph $\Gamma^*(\mathcal{S}_n(\mathbb{F}_q))$ was defined by Y. Egawa [4], who proved that it is distance-regular and computed its parameters.

5. Geometry of Hermitian Matrices

Let D be a division ring which possesses an involution. Denote the involution of D by $-$, i.e.,

$$- : D \to D$$
$$a \mapsto \bar{a}, \tag{16}$$

is a bijective map which has the following properties: for any $a, b \in D$ we have

$$\overline{a+b} = \bar{a} + \bar{b}, \tag{17}$$

$$\overline{ab} = \bar{b}\bar{a}, \tag{18}$$

and

$$\bar{\bar{a}} = a. \tag{19}$$

Let

$$F = \{a \in D | \bar{a} = a\}. \tag{20}$$

Define the trace map

$$Tr : D \to F$$
$$a \mapsto a + \bar{a} \tag{21}$$

and the norm map

$$N : D \to F$$
$$a \mapsto a\bar{a}. \tag{22}$$

We make the following assumptions:

Assumption I F is a proper subfield of D and is contained in the center of D.

Assumption II   The map Tr is surjective.

We remark that Assumption I excludes the case when D is a field and $-$ is the identity map.

Let n be an integer ≥ 2. An $n \times n$ matrix H over D is called *hermitian* if ${}^t\bar{H} = H$. The *space* of the geometry of hermitian matrices over D, denoted by $\mathcal{H}_n(D)$, is the set of all $n \times n$ hermitian matrices over D, whose elements are called the *points*. The set of transformations of $\mathcal{H}_n(D)$ to itself of the form

$$\mathcal{H}_n(D) \to \mathcal{H}_n(D)$$
$$X \mapsto {}^t\bar{P}XP + H, \tag{23}$$

where $P \in GL_n(D)$ and $H \in \mathcal{H}_n(D)$, forms a *transformation group* of the space $\mathcal{H}_n(D)$, which is denoted by $GH_n(D)$. Let $X_1, X_2 \in \mathcal{H}_n(D)$. When rank $(X_1 - X_2) = 1$ then X_1 and X_2 are said to be *adjacent*. Clearly, the adjacency of two points is an invariant under $GH_n(D)$. Conversely, we have

Fundamental Theorem of the Geometry of Hermitian Matrices. Let D be a division ring which possesses an involution and assume that Assumptions I and II hold. Let n be an integer ≥ 2 and when $n = 2$ we assume that D is a field. Let $\mathcal{A}$ be a bijective map from $\mathcal{H}_n(D)$ to itself and assume that both $\mathcal{A}$ and $\mathcal{A}^{-1}$ preserve the adjacency. Then $\mathcal{A}$ is of the form

$$\mathcal{A}(X) = \alpha {}^t\bar{P}X^\sigma P + H \text{ for all } X \in \mathcal{H}_n(D), \tag{24}$$

where $\alpha \in F^*$, $P \in GL_n(D)$, $H \in \mathcal{H}_n(D)$, and σ is an automorphism of D

which commutes with the involution $-$ of D. If we assume further that the norm map N is bijective, then we can assume that $\alpha = 1$,   Q.E.D.

The above theorem was proved by the author [21, 22] recently. In the proof, besides the maximal sets of rank 1 and rank 2, which were defined in a similar way as those in the geometry of symmetric matrices, the reduced maximal sets of rank 2 are also introduced. The normal form of maximal sets of rank 1 under $GH_n(D)$ is

$$\left\{ \left. \begin{pmatrix} x_{11} & 0 & \cdots & 0 \\ 0 & 0 & \cdots & 0 \\ \vdots & \vdots & & \vdots \\ 0 & 0 & \cdots & 0 \end{pmatrix} \right| x_{11} \in F \right\}, \tag{25}$$

and the normal form of maximal sets of rank 2 under $GH_n(D)$ is

$$\left\{ \left. \begin{pmatrix} x_{11} & x_{12} & \cdots & x_{1n} \\ \overline{x_{12}} & 0 & \cdots & 0 \\ \vdots & \vdots & & \vdots \\ \overline{x_{1n}} & 0 & \cdots & 0 \end{pmatrix} \right| x_{11} \in F, x_{12}, \cdots, x_{1n} \in D \right\}. \tag{26}$$

If $\mathcal{M}$ is a maximal set of rank 1, then there is a unique maximal set $\mathcal{L}$ of rank 2 containing $\mathcal{M}$. For any $\mathcal{M}$ containing the zero matrix 0, $\mathcal{L}$ has an additive group structure with respect to matrix addition, $\mathcal{M}$ is its subgroup, and the set of cosets of $\mathcal{L}$ relative to $\mathcal{M}$ is called a *reduced maximal set of rank* 2. Clearly, the reduced maximal set of rank 2 from $\mathcal{L}$ are all the maximal sets of rank 1 contained in $\mathcal{L}$. Hence, if we assume that $\mathcal{A}(0) = 0$, then $\mathcal{A}$ carries reduced maximal sets of rank 2 to reduced maximal sets of rank 2. The reduced maximal sets of rank 2 are used in the proof of the above theorem when $n \geqslant 3$ as the maximal sets in the proof of the fundamental theorem of the geometry of rectangular matrices. When $n = 2$ and D is a field, the theorem can be proved by studying three maximal sets of rank 1 which have a nonempty intersection [22]. The case when $n = 2$ and D is not a field still remains open.

When $D = \mathbb{C}$ and $\mathcal{A}$ satisfies some other conditions, the above theorem was proved by L. K. Hua [5] in 1945. When $D = \mathbb{F}_q$, it was proved by A. A. Ivanov and S. V. Shpectorov [14] in 1991.

The above theorem has also applications to algebra [22] and geometry [23], and can also be interpreted as a theorem on graph automorphisms [22].

References

[1] A. E. Brouwer, A. M. Cohen and A. Neumain, Distance-regular Graphs, Springer-Verlag, 1989.

[2] W. L. Chow, On the geometry of algebraic homogeneous spaces, Ann. of Math. (2), **50** (1949), 32–67.

[3] S. Deng and Q. Li, On the affine geometry of algebraic homogeneous spaces, Acta Math. Sinica, **15**(1965), 651–663 (in Chinese). English translation: Chinese Mathematics, **7** (1965), 387–391.

[4] Y. Egawa, Association schemes of quadratic forms, J. Combin. Theory Ser. A, **38**(1985), 1–14.

[5] L. K. Hua, Geometries of matrices I. Generalizations of van Staudt's theorem, Trans. Amer. Math. Soc., **57** (1945), 441–481.

[6] L. K. Hua, Geometries of matrices I_1. Arithmetical construction, Trans. Amer. Math. Soc., **57** (1945), 482–490.

[7] L. K. Hua, Geometries of symmetric matrices over the real field I, Dokl. Akad. Nauk. SSSR(N. S), **53** (1946), 95–97.

[8] L. K. Hua, Geometries of symmetric matrices over the real field II, Dokl. Akad. Nauk. SSSR(N. S.), **53** (1946), 195–196.

[9] L. K. Hua, Geometries of matrices II. Study of involutions in the geometry of symmetric matrices, Trans. Amer. Math. Soc., **61** (1947), 193–228.

[10] L. K. Hua, Geometries of matrices III. Fundamental theorems in the geometries of symmetric matrices, Trans. Amer. Math. Soc., **61** (1947), 228–255.

[11] L. K. Hua, Geometries of symmetric matrices over any field with characteristic other than two, Ann. of Math., **50** (1949), 8–31.

[12] L. K. Hua, A theorem on matrices over a sfield and its applications, Acta Math. Sinica, **1** (1951), 109–163.

[13] L. K. Hua and Z. Wan, Classical Groups, Shanghai Science and Technology Press, Shanghai, 1963 (in Chinese).

[14] A. A. Ivanov and S. V. Shpectorov, A characterization of the association schemes of Hermitian forms, J. Math. Soc. Japan, **43** (1991), 25–48.

[15] M. Liu, A proof of the fundamental theorem of projective geometry of symmetric matrices, Shuxue Jinzhan, **8** (1965), 283–292 (in Chinese).

[16] M. Liu, Geometry of alternate matrices, Acta Math. Sinica, **16** (1966), 104–135 (in Chinese). English translation: Chinese Mathematics, **8** (1966), 108–143.

[17] D. Pei, A proof of the fundamental theorem of the projective geometry of rectangular matrices, A Collection of Papers Celebrating the Fifth Anniversary of the Chinese University of Science and Technology, 99–110 (in Chinese).

[18] Z. Wan, Geometry of symmetric matrices and its applications I, Algebra Colloq., **1** (1994), 97–120.

[19] Z. Wan, Geometry of symmetric matrices and its applications II, Algebra Colloq., **1** (1994), 201–224.

[20] Z. Wan, The Graph of 3×3 binary symmetric matrices, Northeast. Math. J., **11** (1995), 1–2.

[21] Z. Wan, Geometry of hermitian matrices and its applications I, Algebra Colloq., **2** (1995), 167–192.

[22] Z. Wan, Geometry of hermitian matrices and its applications II, to appear in Algebra Colloq.

[23] Z. Wan, Geometry of Matrices, in preparation.

[24] Z. Wan and Y. Wang, Discussion on "A theorem on matrices over a sfield and its applications", Shuxue Jinzhan, **5** (1962), 325–332 (in Chinese).

Institute of Systems Science
Chinese Academy of Science
Beijing, China
and
Department of Information Theory
Lund University
Lund, Sweden

On the Uniqueness of the Leech Lattice

Zhexian Wan

It has been found that there is an error in Venkov's proof of the uniqueness of the Leech lattice. A construction of neighbours of even unimodular lattices is studied and is used to modify Venkov's proof so that the error is corrected.

©1997 Academic Press Limited

1. Introduction

The Leech lattice was introduced by Leech [4] in 1964 and its uniqueness was proved by Conway [1] in 1969. Then, using a construction of neighbours of unimodular lattices and the fact that there is a unique (up to isomorphism) 24-dimensional even unimodular lattice the roots of which generate the root lattice of type $24A_1$, Venkov [5] gave another proof in 1978. His paper [5] was included as a chapter in Conway and Sloane [2] in 1988 and his proof was reproduced in Ebeling [3] in 1994. However, there is an error in Venkov's proof [5], which will be pointed out explicitly at the end of this paper. Now, properties of a construction of neighbours of even unimodular lattices are exhibited and are used to modify Venkov's proof so that the error is corrected.

2. Construction of Neighbours

Two lattices L and L' in $\mathbb{R}^n$ are called *neighbours* if their intersection $L \cap L'$ has index 2 in each of them.

Proposition 1 *Let L be a unimodular lattice. Then:*

(a) *Let $u \in L$, $u/2 \notin L$ and $u^2/4 \in \mathbb{Z}$. Define $L_u = \{x \in L | x \cdot u \equiv 0 \pmod{2}\}$ and $L^u = L_u \cup (u/2 + L_u)$. Then L^u is also a unimodular lattice, $L \cap L^u = L_u$,* and L and L^u are neighbours.

(b) Any neighbour L' of L arises in the way of (a), i.e. $L' = L^u$, where $u \in L$, $u/2 \notin L$ and $u^2/4 \in \mathbb{Z}$, iff L' is integral.

(c) Let $u, u' \in L, u/2, u'/2 \notin L$ and $u^2/4, u'^2/4 \in \mathbb{Z}$. If $L^u = L^{u'}$, then $L_u = L_{u'}$ and $u/2 \equiv u'/2 (\bmod L_u)$.

Proof (a) This assertion can be found in [2, Chapter 17], but no proof was given there. For completeness, we give a proof of (a).

We assert that $L_u \neq L$. Suppose that $L_u = L$; then $x \cdot u \equiv 0 \pmod 2$ for all $x \in L$; then $x \cdot (u/2) \in \mathbb{Z}$ for all $x \in L$, which implies $u/2 \in L^*$, where L^* denotes the dual lattice of L. Since L is unimodular, $L^* = L$. Then $u/2 \in L$, which is a contradiction. Define $L'_u = \{x \in L | x \cdot u \equiv 1 \pmod 2\}$; then $L = L_u \cup L'_u$ is the coset decomposition of L relative to L_u, $|L/L_u| = 2$ and L_u is a lattice in $\mathbb{R}^n$. Clearly, $L^u = L_u \cup (u/2 + L_u)$ is an integral lattice in $\mathbb{R}^n$ containing L_u, $|L^u/L_u| = 2$ and $L \cap L^u = L_u$. Therefore L^u is a unimodular lattice in $\mathbb{R}^n$, and L and L^u are neighbours.

(b) The 'only if' part follows from (a). Now we are going to prove the 'if' part. Let L' be an integral lattice and L and L' be neighbours. Let $L_d = L \cap L'$. Then $|L/L_d| = |L'/L_d| = 2$ and L' can be written as $L' = L_d \cup (v + L_d)$, where $v \in L'$ and $v \notin L_d$. Since $L \cap L' = L_d, v \notin L$. Clearly, $u = 2v \in L_d$ and, hence, $u \in L, u/2 = v \notin L$ and $u^2/4 = v^2 \in \mathbb{Z}$, since L' is integral. Define L_u as above. As in (a) we can prove that $L_u \neq L$ and $|L/L_u| = 2$. We assert that $L_u = L_d$. For any $x \in L_d$, we have $x \in L'$. Since L' is integral, $x \cdot v \in \mathbb{Z}$, which implies that $x \cdot u \equiv 0 \pmod 2$, i.e. $x \in L_u$. Therefore $L_d \subseteq L_u$. Since $|L/L_d| = |L/L_u| = 2$, we have $L_d = L_u$. Therefore $L' = L_u \cup (u/2 + L_u)$.

(c) Assume that $L^u = L^{u'}$. If $L_u \neq L_{u'}$, there is a $w \in L_u$ but $w \notin L_{u'}$, or a $w \notin L_u$ but $w \in L_{u'}$. Consider the first case. (The other case can be treated in a similar way.) Then $w \in u'/2 + L_{u'}$, and we may write $w = u'/2 + w'$, where $w' \in L_{u'}$. We have $w \cdot u' = u'^2/2 + w' \cdot u' \equiv 0 \pmod 2$, which implies that $w \in L_{u'}$, a contradiction. Therefore $L_u = L_{u'}$. Then $u/2 + L_u = u'/2 + L_u$

and, hence, $u/2 \equiv u'/2 \pmod{L_u}$. □

Proposition 2 *Let L be an even unimodular lattice. Then*:

(a) *Let $u \in L$, $u/2 \notin L$ and $u^2/4 \in \mathbb{Z}$. Then L^u is an even unimodular lattice iff $u^2/8 \in \mathbb{Z}$.*

(b) *Let $u, u' \in L, u/2, u'/2 \notin L, u^2/8, u'^2/8 \in \mathbb{Z}$, and $u/2 \equiv u'/2 \pmod{L}$. Then $L^u = L^{u'}$ and $u/2 \equiv u'/2 \pmod{L_u}$.*

Proof (a) By Proposition 1(a), we know that L^u is a unimodular lattice. For any $w \in L_u$, we have $w \cdot u \equiv 0 \pmod 2$. Since L is even and $L \supset L_u$, $w^2 \in 2\mathbb{Z}$. Therefore $(u/2+w)^2 = u^2/4 + u \cdot w + w^2 \in 2\mathbb{Z}$ iff $u^2/4 \in 2\mathbb{Z}$, i.e. $u^2/8 \in \mathbb{Z}$.

(b) Clearly, $u = u' + 2w$, where $w \in L$. Thus $x \cdot u \equiv x \cdot u' \pmod 2$ for all $x \in L$, which implies that $L_u = L_{u'}$. From $u/2 = u'/2 + w$, we deduce that $u^2/4 = u'^2/4 + u' \cdot w + w^2$. By hypothesis, $u^2/4, u'^2/4$ and w^2 are all even. It follows that $u' \cdot w \equiv 0 \pmod 2$, i.e. $w \in L_{u'} = L_u$. Then $u/2 \equiv u'/2 \pmod{L_u}$ and $u/2 + L_u = u'/2 + L_{u'}$. Therefore $L^u = L^{u'}$. □

Example 1 Let $\widetilde{G}$ be the extended binary Golay code, which is a doubly even self-dual binary linear [24, 12, 8]-code. Let

$$L_{\widetilde{G}} = \left\{ \frac{1}{\sqrt{2}}(c+2y) \,\Big|\, c \in \widetilde{G} \text{ and } y \in \mathbb{Z}^{24} \right\},$$

where the c's are regarded as 24-dimensional vectors, the components of which are real numbers 0 and 1, not elements from $\mathbb{F}_2$. It is known that $L_{\widetilde{G}}$ is an even unimodular lattice in $\mathbb{R}^{24}$ the roots (i.e. vectors of square length 2) of which generate the root lattice of type $24A_1$. Let

$$\Lambda_{24} = \left\{ \frac{1}{\sqrt{2}}(c+2y) \,\Big|\, c \in \widetilde{G}, y \in \mathbb{Z}^{24}, \text{and } \sum_{i=1}^{24} y_i \equiv 0 \pmod 2 \right\}$$

$$\cup \left\{ \frac{1}{\sqrt{2}}\left(\frac{1}{2}1^{24} + c + 2z\right) \,\Big|\, c \in \widetilde{G}, z \in \mathbb{Z}^{24}, \text{and } \sum_{i=1}^{24} z_i \equiv 1 \pmod 2 \right\},$$

where 1^{24} is the 24-dimensional all-1 vector. It is also known that Λ_{24} is the Leech lattice, which is an even unimodular lattice in $\mathbb{R}^{24}$ without roots. Let $u = (1/\sqrt{2})(-3, 1^{23})$, where 1^{23} represents 23 1's. It is easy to see that

$u \in L_{\widetilde{G}}, u/2 \notin L_{\widetilde{G}}, u^2/8 \in \mathbb{Z}$,

$$(L_{\widetilde{G}})_u = \left\{ \frac{1}{\sqrt{2}}(c + 2y) \,\Big|\, c \in \widetilde{G}, y \in \mathbb{Z}^{24}, \text{and} \sum_{i=1}^{24} y_i \equiv 0 (\text{mod } 2) \right\},$$

$$(L_{\widetilde{G}})'_u = \left\{ \frac{1}{\sqrt{2}}(c + 2z) \,\Big|\, c \in \widetilde{G}, z \in \mathbb{Z}^{24}, \text{and} \sum_{i=1}^{24} z_i \equiv 1 (\text{mod } 2) \right\},$$

$$L_{\widetilde{G}} = (L_{\widetilde{G}})_u \cup (L_{\widetilde{G}})'_u$$

and

$$\frac{u}{2} + (L_{\widetilde{G}})_u = \left\{ \frac{1}{\sqrt{2}} \left(\frac{1}{2} 1^{24} + c + 2z \right) \,\Big|\, c \in \widetilde{G}, z \in \mathbb{Z}^{24}, \text{and} \sum_{i=1}^{24} z_i \equiv 1 (\text{mod } 2) \right\}.$$

Therefore

$$\Lambda_{24} = (L_{\widetilde{G}})^u = (L_{\widetilde{G}})_u \cup (u/2 + (L_{\widetilde{G}})_u).$$

Example 2 Let $L = \mathbb{Z}^2$ and $L' = \frac{1}{2}\mathbb{Z} \times 2\mathbb{Z}$. Then $L \cap L' = \mathbb{Z} \times 2\mathbb{Z}$. Clearly, $|L/L \cap L'| = |L'/L \cap L'| = 2$. Hence L and L' are neighbours. Since L' is not an integral lattice, it cannot be expressed as $L^u = L_u \cup (u/2 + L_u)$, where $u \in L, u/2 \notin L, u^2/4 \in \mathbb{Z}$ and $L_u = \{x \in L | x \cdot u \equiv 0 \ (\text{mod } 2)\}$.

This example shows that the statement 'All neighbours of a unimodular lattice arise in the way of (a)', which was stated in [2, Chapter 17] is not correct.

Example 3 In Example 1, let $u' = (1/\sqrt{2})1^{24}$. Then $u' \in L_{\widetilde{G}}, u'/2 \notin L_{\widetilde{G}}, u'^2/4 = 3 \in \mathbb{Z}$ and $u'^2/8 \notin \mathbb{Z}$. By Proposition 2(a), $(L_{\widetilde{G}})^{u'}$ is not even. Therefore $(L_{\widetilde{G}})^u \neq (L_{\widetilde{G}})^{u'}$. Clearly, $u/2 - u'/2 = (1/\sqrt{2})(-2, 0^{23}) \in L_{\widetilde{G}}$.

This example shows that the statement 'for a unimodular (or an even unimodular) lattice L, and for $u, u' \in L, u/2, u'/2 \notin L$, and $u^2/4, u'^2/4 \in \mathbb{Z}, L^u = L^{u'}$ iff $u/2 \equiv u'/2 \ (\text{mod } L)$', which was stated in [2, Chapter 17] (or [3, Chapter 4] respectively), is not correct.

3. The Uniqueness of the Leech Lattice

Theorem 3 (Conway) *There is a unique (up to isomorphism) 24-dimensional even unimodular lattice without roots.*

Proof First we follow Venkov's proof. Let L be a 24-dimensional even unimodular lattice without roots and let $\theta_L(z)$ be its theta function. Let Λ_{24} be the Leech lattice and let $\theta_{\Lambda_{24}}(z)$ be its theta function. Both $\theta_L(z)$ and $\theta_{\Lambda_{24}}(z)$ are modular forms of weight 12. Expanding both of them into power series in q, where $q = e^{2\pi i z}$, their constant terms are both equal to 1 and their coefficients of q^2 are both 0. Therefore $\theta_L(z) = \theta_{\Lambda_{24}}(z)$. Since there is a vector in Λ_{24}, say $(1/\sqrt{2})(1^8, 0^8, 1^8)$, of square length 8, the coefficient of q^4 in $\theta_{\Lambda_{24}}(z)$ is > 0. So, the coefficient of q^4 in $\theta_L(z)$ is > 0, and, hence, there is a vector $u \in L$ such that $u^2 = 8$. Since $(u/2)^2 = 2$ and L has no roots, $u/2 \notin L$. By Propositions 1(a) and 2(a), $L^u = L_u \cup (u/2 + L_u)$ is an even unimodular lattice in $\mathbb{R}^{24}$. Since $u/2 \in L^u$, L^u has roots.

We prove that the roots in L^u generate a root lattice of type $24A_1$. It is sufficient to show that for any two roots $x, y \in L^u$ and $x \neq \pm y$, we have $x \cdot y = 0$. We have $x^2 = y^2 = 2$. Assume that $x \cdot y \neq 0$. Then either $x - y$ or $x + y$ is a root, and both belong to L_u. But L_u has no root, since $L_u \subset L$, a contradiction. Therefore $L^u \simeq L_{\widetilde{G}}$.

Let $e_i = (1/\sqrt{2})\epsilon_i$, where $\epsilon_i = (0^{i-1} \ 1 \ 0^{24-i})$, $i = 1, \cdots, 24$. Then $e_i \cdot e_j = 0$ for $i \neq j$, $e_i^2 = \dfrac{1}{2}$, and $e_1, \cdots, e_{24}$ is a basis of $\mathbb{R}^n$. Clearly, $2e_i \in L_{\widetilde{G}}$. Under the isomorphism $L^u \simeq L_{\widetilde{G}}$, assume that $v_i \mapsto 2e_i$, $i = 1, \cdots, 24$. Then $v_1, \cdots, v_{24} \in L^u$ with $v_i \cdot v_j = 0$ for $i \neq j$ and $v_i^2 = 2$, $i = 1, \cdots, 24$, and $v_1, \cdots, v_{24}$ is a basis of $\mathbb{R}^n$.

Now we deviate from Venkov's proof. Since L^u and L are neighbours, by Propositions 1(b) and 2(a), there is a $v \in L^u$ with $v/2 \notin L^u$, $v^2/8 \in \mathbb{Z}$, such that $L = (L^u)^v = (L^u)_v \cup (v/2 + (L^u)_v)$. Assume that $v = \sum_{i=1}^{24} \dfrac{1}{2} m_i v_i$, $m_i \in \mathbb{Z}$. Since L^u is integral, $v_i \cdot v \in \mathbb{Z} (i = 1, \cdots, 24)$, so $m_i \in \mathbb{Z}$. We assert that all m_i are odd. Suppose that m_i is even for some i. Then $v_i \cdot v = m_i \equiv 0 \pmod{2}$, i.e. $v_i \in (L^u)_v \subset L$, which contradicts the assertion that L has no roots.

For any $i > 1$, we have

$$(v_1 + v_i) \cdot v = m_1 + m_i \equiv 0 \pmod{2}.$$

It follows that $v_1 + v_i \in (L^u)_v$. Clearly,

$$m_i = 4q_i + \eta_i, \quad \text{where } q_i \in \mathbb{Z} \text{ and } \eta_i = \pm 1, \ i = 2, \cdots, 24.$$

Then $v' = v - \sum_{i=2}^{24} q_i(2v_1 + 2v_i)$ is of the form $\frac{1}{2}(n_1 v_1 = \eta_2 v_2 + \cdots + \eta_{24} v_{24})$, where n_1 is odd. Clearly, $v' \in L^u$, $v'/2 \notin L^u$, $v'^2/8 \in \mathbb{Z}$ and $v'/2 \equiv v/2 \pmod{(L^u)_v}$. By Proposition 2(b), $(L^u)^v = (L^u)^{v'}$. Therefore, we can assume that

$$v = \frac{1}{2}(\eta_1 v_1 + \eta_2 v_2 + \cdots + \eta_{24} v_{24}).$$

Similarly, $2v_1 \in (L^u)_v$ and we can assume that $n_1 \in \{\pm 1, \pm 3\}$. If $n_1 = \pm 1$, then $v^2 = 12$, i.e. $v^2/8 \notin \mathbb{Z}$. Therefore $n_1 = \pm 3$. If $\eta_i < 0$ for some i, let σ_{v_i} be the reflection determined by the root v_i. Then

$$\sigma_{v_i}(v_j) = v_j \quad \text{for } j \neq i, \quad \sigma_{v_i}(v_i) = -v_i,$$
$$\sigma_{v_i}(L^u) = L^u$$

and

$$\sigma_{v_i}(L^u)_v = (L^u)_{\sigma_{v_i}(v)}.$$

Clearly,

$$(L^u)^{\sigma_{v_i}(v)} = (L^u)_{\sigma_{v_i}(v)} \cup \left(\frac{\sigma_i(v)}{2} + (L^u)_{\sigma_{v_i}(v)}\right) \simeq (L^u)_v \cup \left(\frac{v}{2} + (L^u)_v\right) = (L^u)^v.$$

Hence we can assume that $\eta_i = 1$ for all $i > 1$. Similarly, we can assume that $n_1 = -3$. Therefore we can assume that $v = \frac{1}{2}(-3v_1 + v_2 + \cdots + v_{24})$, which satisfies the condition $v \in L^u, v/2 \notin L^u$ and $v^2/8 \in \mathbb{Z}$. By Example 1, $(L_{\widetilde{G}})^u = \Lambda_{24}$, where $u = (1/\sqrt{2})(-3, 1^{23})$; therefore we also have $(L^u)^v \simeq \Lambda_{24}$. □

Remark In Venkov's proof [5], the case $v = \frac{1}{2}(-3v_1 + v_2 + \cdots + v_{24})$ is missing and only the incorrect case $v = \frac{1}{2}(v_1 + v_2 + \cdots + v_{24})$ appears. But $v = \frac{1}{2}(v_1 + v_2 + \cdots + v_{24})$ does not satisfy $v^2/8 \in \mathbb{Z}$. By Proposition 2(a), $(L^u)^v$ is not even, so $(L^u)^v \neq \Lambda_{24}$.

References

[1] J. H. Conway, A characterization of Leech's lattice. *Invent. Math.* **121** (1969), 119–133.

[2] J. H. Conway and N. J. A. Sloane, *Sphere Packings, Lattices and Groups*, Springer, New York, 1982, 2nd edn 1993.

[3] W. Ebeling, *Lattices and Codes*, Vieweg, Wiesbaden, 1994.

[4] J. Leech, Some sphere packings in higher space, *Can. J. Math.*, **16** (1964), 657–682.

[5] B. B. Venkov, On the classification of integral even unimodular 24-dimensional quadratic forms, *Trudy Matematischeskogo Instituta imeni V. A. Steklova*, **148** (1978), 65–76 (in Russian). English translation: *Proc. Steklov Inst. Math.*, **4** (1980), 63–74.

Received 10 January 1996 and accepted in revised form 2 April 1996
Zhexian Wan
Institute of Systems Science,
Chinese Academy of Science,
Beijing 100080, China
and
Department of Information Theory,
Lund University,
Box 118, S-221 00
Lund, Sweden

Symplectic Graphs and Their Automorphisms

Zhongming Tang[a,*], Zhexian Wan[b]

a Department of Mathematics, Suzhou University, Suzhou 215006, PR China
b Academy of Mathematics and System Sciences, Chinese Academy of Science, Beijing 100080, PR China

Received 10 May 2004; accepted 16 August 2004
Available online 18 September 2004

Abstract A new family of strongly regular graphs, called the general symplectic graphs $Sp(2\nu, q)$, associated with nonsingular alternate matrices is introduced. Their parameters as strongly regular graphs, their chromatic numbers as well as their groups of graph automorphisms are determined.
© 2004 Elsevier Ltd. All rights reserved.

Keywords Symplectic graphs; Chromatic numbers; Graph automorphisms

1. Introduction

Let $\mathbb{F}_q$ be a finite field of any characteristic and $\nu \geqslant 1$ be an integer. Let

$$\mathbb{F}_q^{(2\nu)} = \{(a_1, \cdots, a_{2\nu}) : a_i \in \mathbb{F}_q, i = 1, \cdots, 2\nu\}.$$

be the 2ν-dimensional row vector space over $\mathbb{F}_q$. For any $\alpha_1, \cdots, \alpha_n \in \mathbb{F}_q^{(2\nu)}$, we denote the subspace of $\mathbb{F}_q^{(2\nu)}$ generated by $\alpha_1, \cdots, \alpha_n$ by $[\alpha_1, \cdots, \alpha_n]$. Thus, if $\alpha = (a_1, \cdots, a_{2\nu}) \neq 0 \in \mathbb{F}_q^{(2\nu)}$ then $[\alpha]$, which is also denoted by $[a_1, \cdots, a_{2\nu}]$, is a one-dimensional subspace of $\mathbb{F}_q^{(2\nu)}$ and $[\alpha] = [k\alpha]$ for any $k \in \mathbb{F}_q^* = \mathbb{F}_q \backslash \{0\}$.

---

* Corresponding author.
E-mail addresses: zmtang@suda.edu.cn (Z. Tang), wan@amss.ac.cn (Z.-x. Wan).

Let K be a $2\nu \times 2\nu$ nonsingular alternate matrix over $\mathbb{F}_q$. The *symplectic graph* relative to K over $\mathbb{F}_q$ is the graph with the set of one-dimensional subspaces of $\mathbb{F}_q^{(2\nu)}$ as its vertex set and with the adjacency defined by

$$[\alpha] \sim [\beta] \text{ if and only if } \alpha K^t \beta \neq 0, \text{ for any } \alpha \neq 0, \beta \neq 0 \in \mathbb{F}_q^{(2\nu)},$$

where $[\alpha] \sim [\beta]$ means that $[\alpha]$ and $[\beta]$ are adjacent. Since any two $2\nu \times 2\nu$ nonsingular alternate matrices over $\mathbb{F}_q$ are cogredient, any two symplectic graphs relative to two different $2\nu \times 2\nu$ nonsingular alternate matrices over $\mathbb{F}_q$ are isomorphic. Thus we can assume that

$$K = \begin{pmatrix} 0 & 1 & & & & \\ -1 & 0 & & & & \\ & & 0 & 1 & & \\ & & -1 & 0 & & \\ & & & & \ddots & \\ & & & & & 0 & 1 \\ & & & & & -1 & 0 \end{pmatrix}_{2\nu \times 2\nu}$$

and consider only the symplectic graph relative to the above K over $\mathbb{F}_q$, which will be denoted by $Sp(2\nu, q)$.

When $q = 2$, the special case $Sp(2\nu, 2)$ of the graph $Sp(2\nu, q)$ was studied previously by Rotman [4], Rotman and Weichsel [5], Godsil and Royle [2,3] etc. In the present paper we study the general case $Sp(2\nu, q)$. In Section 2, we show that $Sp(2\nu, q)$ is strongly regular and compute its parameters. We also prove that the chromatic number of $Sp(2\nu, q)$ is $q^\nu + 1$. Section 3 is devoted to discussing the group of automorphisms Aut $(Sp(2\nu, q))$ of the graph. The structure of this group depends on q and ν. When $q = 2$, Aut $(Sp(2\nu, 2))$ is isomorphic to the symplectic group of degree 2ν over $\mathbb{F}_2$. When $q > 2$, Aut $(Sp(2\nu, q))$ is the product of two subgroups which are identified clearly (cf. Theorem 3.4).

2. Strong regularity and chromatic numbers of symplectic graphs

For any subspace V of $\mathbb{F}_q^{(2\nu)}$, we denote the subspace of $\mathbb{F}_q^{(2\nu)}$ formed by all $\beta \in \mathbb{F}_q^{(2\nu)}$ such that $\alpha K^t \beta = 0$ for all $\alpha \in V$ by $V^\perp$. Then $[\alpha] \sim [\beta]$ if and only if $\beta \notin [\alpha]^\perp$.

Denote the vertex set of the graph $Sp(2\nu, q)$ by $V(Sp(2\nu, q))$. We first show that $Sp(2\nu, q)$ is strongly regular.

Theorem 2.1 *$Sp(2\nu, q)$ is a strongly regular graph with parameters*

$$\left(\frac{q^{2\nu}-1}{q-1}, q^{2\nu-1}, q^{2\nu-2}(q-1), q^{2\nu-2}(q-1)\right)$$

and eigenvalues $q^{2\nu-1}$, $q^{\nu-1}$ and $-q^{\nu-1}$.

Proof As $|\mathbb{F}_q^{(2\nu)}| = q^{2\nu}$, it follows that $|V(Sp(2\nu, q))| = \dfrac{q^{2\nu}-1}{q-1}$. For any $[\alpha] \in V(Sp(2\nu, q))$, since $\dim([\alpha]^\perp) = 2\nu - 1$, we see that the degree of $[\alpha]$, which is just the number of one-dimensional subspaces $[\beta]$ such that $\beta \notin [\alpha]^\perp$, is $\dfrac{q^{2\nu}-q^{2\nu-1}}{q-1} = q^{2\nu-1}$.

Let $[\alpha]$, $[\beta]$ be any two different vertices of $Sp(2\nu, q)$ which are adjacent to each other or not. Then $\dim([\alpha, \beta]^\perp) = 2\nu - 2$. Note that a vertex $[\gamma]$ is adjacent to both $[\alpha]$ and $[\beta]$ is equivalent to that $\gamma \notin [\alpha]^\perp \cup [\beta]^\perp$. But

$$|[\alpha]^\perp \cup [\beta]^\perp| = |[\alpha]^\perp| + |[\beta]^\perp| - |[\alpha, \beta]^\perp|.$$

Hence the number of vertices which are adjacent to both $[\alpha]$ and $[\beta]$ is $\dfrac{q^{2\nu}-2q^{2\nu-1}+q^{2\nu-2}}{q-1} = q^{2\nu-2}(q-1)$. Therefore $Sp(2\nu, q)$ is a strongly regular graph with parameter

$$\left(\frac{q^{2\nu}-1}{q-1}, q^{2\nu-1}, q^{2\nu-2}(q-1), q^{2\nu-2}(q-1)\right).$$

By the same arguments as in [3, Section 10.2], we get that the eigenvalues of $Sp(2\nu, q)$ are $q^{2\nu-1}$, $q^{\nu-1}$ and $-q^{\nu-1}$. □

Let $n \geq 2$. We say that a graph X is *n-partite* if there are subsets $X_1, \cdots, X_n$ of the vertex set $V(X)$ of X such that $V(X) = X_1 \cup \cdots \cup X_n$, where $X_i \cap X_j = \varnothing$ for all $i \neq j$, and that there is no edge of X joining two vertices of the same subset. We are going to show that $Sp(2\nu, q)$ is $(q^\nu + 1)$-partite. We need some results about subspaces of $\mathbb{F}_q^{(2\nu)}$. A subspace V of $\mathbb{F}_q^{(2\nu)}$ is called *totally isotropic* if $V \subseteq V^\perp$. Then totally isotropic subspaces of $\mathbb{F}_q^{(2\nu)}$

are of dimension $\leqslant \nu$ and there exist totally isotropic subspaces of dimension ν which are called *maximal totally isotropic subspaces*; cf. [6, Corollary 3.8].

The following lemma is due to Dye [1].

Lemma 2.2 *There exist maximal totally isotropic subspaces $V_i, i = 1, \cdots, q^\nu + 1$, of $\mathbb{F}_q^{(2\nu)}$ such that*

$$\mathbb{F}_q^{(2\nu)} = V_1 \cup \cdots \cup V_{q^\nu+1},$$

where $V_i \cap V_j = \{0\}$ for all $i \neq j$.

Proposition 2.3 $Sp(2\nu, q)$ *is $(q^\nu + 1)$-partite. That is, there exist subsets $X_1, \cdots, X_{q^\nu+1}$ of $V(Sp(2\nu, q))$ such that*

$$V(Sp(2\nu, q)) = X_1 \cup \cdots \cup X_{q^\nu+1},$$

where $X_i \cap X_j = \emptyset$ for all $i \neq j$, and there is no edge of $Sp(2\nu, q)$ joining two vertices of the same subset. Moreover, the subsets $X_1, \cdots, X_{q^\nu+1}$ can be so chosen that for any two distinct indices i and j, every $\alpha \in X_i$ is adjacent to exactly $q^{\nu-1}$ vertices in X_j.

Proof Let $\mathbb{F}_q^{(2\nu)} = V_1 \cup \cdots \cup V_{q^\nu+1}$ as in 2.2. Set $X_i = \{[\alpha] : \alpha \neq 0 \in V_i\}$, $i = 1, \cdots, q^\nu + 1$. Then

$$V(Sp(2\nu, q)) = X_1 \cup \cdots \cup X_{q^\nu+1}, X_i \cap X_j = \emptyset, \text{ for all } i \neq j.$$

As V_i is totally isotropic, we see that there is no edge joining any two vertices in X_i. Thus $Sp(2\nu, q)$ is $(q^\nu + 1)$-partite. For any $i \neq j$, let $[\alpha] \in X_i$. Since V_j is maximal totally isotropic of dimension ν, it follows that $\alpha \notin V_j = V_j^\perp$ and $\dim([\alpha]^\perp \cap V_j) = \dim([\alpha, V_j]^\perp) = \nu - 1$. Note that, for any $[\beta] \in X_j$, $[\beta]$ is adjacent to $[\alpha]$ if and only if $\beta \in V_j \setminus ([\alpha]^\perp \cap V_j)$. Hence the number of vertices in X_j which are adjacent to $[\alpha]$ is $\dfrac{q^\nu - 1}{q - 1} - \dfrac{q^{\nu-1} - 1}{q - 1} = q^{\nu-1}$. □

Now we can compute the chromatic number of $Sp(2\nu, q)$.

Theorem 2.4 $\chi(Sp(2\nu, q)) = q^\nu + 1$.

Proof By 2.3, we see that $\chi(Sp(2\nu, q)) \leqslant q^\nu + 1$. Note that $\chi(Sp(2\nu, q))$ is the minimal n such that $Sp(2\nu, q)$ is n-partite. Suppose that $Sp(2\nu, q)$ is n-partite. Then there exist subsets $Y_1, \cdots, Y_n$ of $V(Sp(2\nu, q))$ such that

$$V(Sp(2\nu,q)) = Y_1 \cup \cdots \cup Y_n, Y_i \cap Y_j = \varnothing, \quad \text{for all } i \neq j,$$

and there is no edge joining any two vertices in the same Y_i for $i = 1, \cdots, n$. We want to show that $n \geqslant q^\nu + 1$. Suppose that $n < q^\nu + 1$. From the above equality, we have $\sum_{i=1}^n |Y_i| = \dfrac{q^{2\nu}-1}{q-1} = \left(\dfrac{q^\nu-1}{q-1}\right)(q^\nu+1)$. Then there exists some i such that $|Y_i| > \dfrac{q^\nu-1}{q-1}$. Let W_i be the subspace of $\mathbb{F}_q^{(2\nu)}$ generated by all α such that $[\alpha] \in Y_i$. Then W_i is a totally isotropic subspace; hence $\dim W_i \leqslant \nu$. It turns out that $|Y_i| \leqslant \dfrac{q^\nu-1}{q-1}$, a contradiction. Hence $\chi(Sp(2\nu,q)) = q^\nu + 1$. □

3. Automorphisms of symplectic graphs

We recall that a $2\nu \times 2\nu$ matrix T is called a *symplectic matrix* (or *generalized symplectic matrix*) of order 2ν over $\mathbb{F}_q$ if $TK^tT = K$ (or $TK^tT = kK$ for some $k \in \mathbb{F}_q^*$, respectively). The set of symplectic matrices (or generalized symplectic matrices) of order 2ν over $\mathbb{F}_q$ forms a group with respect to the matrix multiplication, which is called the *symplectic group* (or *generalized symplectic group*, respectively) of degree 2ν over $\mathbb{F}_q$ and denoted by $Sp_{2\nu}(\mathbb{F}_q)$ (or $GSp_{2\nu}(\mathbb{F}_q)$). The center of $Sp_{2\nu}(\mathbb{F}_q)$ consists of the identity matrix E and $-E$, and the factor group $Sp_{2\nu}(\mathbb{F}_q)/\{E,-E\}$ is called the *projective symplectic group* of degree 2ν over $\mathbb{F}_q$ and denoted by $PSp_{2\nu}(\mathbb{F}_q)$. The center of $GSp_{2\nu}(\mathbb{F}_q)$ consists of all kE, where $k \in \mathbb{F}_q^*$, and the factor group of $GSp_{2\nu}(\mathbb{F}_q)$ with respect to its center is called the *projective generalized symplectic group* of degree 2ν over $\mathbb{F}_q$ and denoted by $PGSp_{2\nu}(\mathbb{F}_q)$. Clearly, $PGSp_{2\nu}(\mathbb{F}_q) \cong PSp_{2\nu}(\mathbb{F}_q)$, and when $q = 2$, $GSp_{2\nu}(\mathbb{F}_2) = Sp_{2\nu}(\mathbb{F}_2)$.

Proposition 3.1 Let T be a $2\nu \times 2\nu$ nonsingular matrix over $\mathbb{F}_q$ and
$$\sigma_T : V(Sp(2\nu,q)) \to V(Sp(2\nu,q))$$
$$[\alpha] \mapsto [\alpha T].$$

Then

(1) $T \in GSp_{2\nu}(\mathbb{F}_q)$ if and only if $\sigma_T \in \text{Aut}(Sp(2\nu,q))$. In particular, when $q = 2$, $T \in Sp_{2\nu}(\mathbb{F}_2)$ if and only if $\sigma_T \in \text{Aut}(Sp(2\nu,2))$.

(2) *For any $T_1, T_2 \in GSp_{2\nu}(\mathbb{F}_q), \sigma_{T_1} = \sigma_{T_2}$ if and only if $T_1 = kT_2$ for some $k \in \mathbb{F}_q$.*

Proof It is clear that σ_T is a one-one correspondence from $V(Sp(2\nu, q))$ to itself.

(1) First assume $T \in GSp_{2\nu}(\mathbb{F}_q)$. Then $TK^tT = kK$ for some $k \in \mathbb{F}_q^*$. For any $[\alpha], [\beta] \in V(Sp(2\nu, q))$, since $\alpha K^t \beta = k^{-1}(\alpha T)K^t(\beta T)$, $[\alpha] \sim [\beta]$ if and only if $\sigma_T([\alpha]) \sim \sigma_T([\beta])$; hence $\sigma_T \in \mathrm{Aut}(Sp(2\nu, q))$.

Conversely, assume $\sigma_T \in \mathrm{Aut}(Sp(2\nu, q))$. Then, for any $\alpha, \beta \neq 0 \in \mathbb{F}_q^{(2\nu)}$, $\alpha K^t \beta = 0$ if and only if $\alpha(TK^tT)^t \beta = 0$. Hence, for any $\alpha \neq 0 \in \mathbb{F}_q^{(2\nu)}$, the two systems of linear equations $(\alpha K)^t X = 0$, $(\alpha T K^t T)^t X = 0$ have the same solutions. But rank $(\alpha K) = \mathrm{rank}(\alpha TK^tT) = 1$; we see that $\alpha K = k(\alpha TK^tT)$ for some $k \in \mathbb{F}_q^*$, which depends on α. Take $\alpha = (1, 0, \cdots, 0)$, $(0, 1, \cdots, 0), \cdots, (0, 0, \cdots, 1)$; we get that $K = \mathrm{diag}(k_1, k_2, \cdots, k_{2\nu})TK^tT$, for some $k_1, k_2, \cdots, k_{2\nu} \in \mathbb{F}_q^*$. Take $\alpha = (1, 1, \cdots, 1)$; we see that $k_1 = k_2 = \cdots = k_{2\nu}$, hence $K = k_1 TK^tT$.

(2) It is clear that $\sigma_{T_1} = \sigma_{T_2}$ if $T_1 = kT_2$ for some $k \in \mathbb{F}_q^*$. Conversely, suppose that $\sigma_{T_1} = \sigma_{T_2}$. Then, for any $\alpha \neq 0 \in \mathbb{F}_q^{(2\nu)}$, $\alpha T_1 = k\alpha T_2$ for some $k \in \mathbb{F}_q^*$. Take $\alpha = (1, 0, \cdots, 0), (0, 1, \cdots, 0)$, and so on as above; we see that $T_1 = kT_2$ for some $k \in \mathbb{F}_q^*$. □

By Proposition 3.1, every generalized symplectic matrix in $GSp_{2\nu}(\mathbb{F}_q)$ induces an automorphism of $Sp(2\nu, q)$ and two generalized symplectic matrices T_1 and T_2 induce the same automorphism of $Sp(2\nu, q)$ if and only if $T_1 = kT_2$ for some $k \in \mathbb{F}_q$. Thus $PSp_{2\nu}(\mathbb{F}_q)$ can be regarded as a subgroup of $\mathrm{Aut}(Sp(2\nu, q))$.

Proposition 3.2 *$Sp(2\nu, q)$ is vertex transitive and edge transitive.*

Proof For any $[\alpha], [\beta] \in V(Sp(2\nu, q))$, there exists $T \in Sp_{2\nu}(\mathbb{F}_q)$ such that $\alpha T = \beta$ by [6, Lemma 3.11]. Then $\sigma_T \in \mathrm{Aut}(Sp(2\nu, q))$ such that $\sigma_T([\alpha]) = [\beta]$. Hence $Sp(2\nu, q)$ is vertex transitive.

Let $[\alpha_1], [\alpha_2], [\beta_1], [\beta_2] \in V(Sp(2\nu, q))$ such that $[\alpha_1] \sim [\alpha_2]$ and $[\beta_1] \sim [\beta_2]$. We may assume that $\alpha_1 K^t \alpha_2 = \beta_1 K^t \beta_2$. Then, by [6, Lemma 3.11] again, there exists $T \in Sp_{2\nu}(\mathbb{F}_q)$ such that $\alpha_1 T = \beta_1$ and $\alpha_2 T = \beta_2$. Then $\sigma_T \in \text{Aut}(Sp(2\nu, q))$ such that $\sigma_T([\alpha_1]) = [\beta_1]$ and $\sigma_T([\alpha_2]) = [\beta_2]$. Hence $Sp(2\nu, q)$ is edge transitive. □

When $q = 2$, we have the following:

Proposition 3.3 $\text{Aut}(Sp(2\nu, 2)) \cong Sp_{2\nu}(\mathbb{F}_2)$.

Proof Let

$$\sigma : Sp_{2\nu}(\mathbb{F}_2) \to \text{Aut}(Sp(2\nu, 2))$$
$$T \mapsto \sigma_T.$$

Then, by Proposition 3.1, σ is an injection. Clearly, σ preserves the operation. It remains to show that, for any $\tau \in \text{Aut}(Sp(2\nu, 2))$, there exists a $T \in Sp_{2\nu}(\mathbb{F}_2)$ such that $\tau = \sigma_T$.

Note that, for any $\alpha \neq 0 \in \mathbb{F}_q^{(2\nu)}$, we have that $[\alpha] = \{0, \alpha\}$. We will denote the uniquely defined element $\tau([\alpha]) \backslash \{0\}$ by $\tau(\alpha)$ and set $\tau(0) = 0$. Then from $\tau \in \text{Aut}(Sp(2\nu, 2))$ we see that $\alpha K^t \beta = \tau(\alpha) K^t(\tau(\beta))$ for any $\alpha, \beta \in \mathbb{F}_2^{(2\nu)}$ (not necessarily non-zero). Fix any $\alpha \in \mathbb{F}_q^{(2\nu)}$. Let $\beta_1, \beta_2 \in \mathbb{F}_2^{(2\nu)}$. Then

$$\alpha K^t \beta_1 = \tau(\alpha) K^t(\tau(\beta_1)),$$

$$\alpha K^t \beta_2 = \tau(\alpha) K^t(\tau(\beta_2)).$$

Thus

$$\alpha K^t (\beta_1 + \beta_2) = \tau(\alpha) K^t(\tau(\beta_1) + \tau(\beta_2)).$$

But

$$\alpha K^t (\beta_1 + \beta_2) = \tau(\alpha) K^t(\tau(\beta_1 + \beta_2));$$

hence

$$\tau(\alpha) K^t (\tau(\beta_1 + \beta_2) + \tau(\beta_1) + \tau(\beta_2)) = 0.$$

This is true for any $\alpha \in \mathbb{F}_2^{(2\nu)}$; it follows that $\tau(\beta_1 + \beta_2) + \tau(\beta_1) + \tau(\beta_2) = 0$, i.e., $\tau(\beta_1 + \beta_2) = \tau(\beta_1) + \tau(\beta_2)$. Set

$$T = \begin{pmatrix} \tau(1,0,\cdots,0) \\ \tau(0,1,\cdots,0) \\ \vdots \\ \tau(0,0,\cdots,1) \end{pmatrix}.$$

Then $\tau(\alpha) = \alpha T$ for any $\alpha \in \mathbb{F}_2^{(2\nu)}$. Thus T is nonsingular. By Proposition 3.1 $T \in Sp_{2\nu}(\mathbb{F}_2)$ and $\tau = \sigma_T$ as required. □

From now on, we assume that $q > 2$. In $\mathbb{F}_q^{(2\nu)}$, let us set

$$e_1 = (1,0,0,0,\cdots,0,0),$$
$$f_1 = (0,1,0,0,\cdots,0,0),$$
$$e_2 = (0,0,1,0,\cdots,0,0),$$
$$f_2 = (0,0,0,1,\cdots,0,0),$$
$$\vdots$$
$$e_\nu = (0,0,0,0,\cdots,1,0),$$
$$f_\nu = (0,0,0,0,\cdots,0,1).$$

Then $e_i, f_i, i = 1,\cdots,\nu$, form a basis of $\mathbb{F}_q^{(2\nu)}$ and $e_i K^t f_i = 1, e_i K^t e_j = 0$, $f_i K^t f_j = 0, i,j = 1,\cdots,\nu$, and $e_i K^t f_j = 0, i \neq j, i,j = 1,\cdots,\nu$.

In order to describe $\mathrm{Aut}(Sp(2\nu,q))$ for any prime power q, we need a definition from group theory. Let φ be the natural action of $\mathrm{Aut}(\mathbb{F}_q)$ on the group $\mathbb{F}_q^* \times \cdots \times \mathbb{F}_q^*$ (ν in number) defined by

$$\varphi(\pi)((k_1,\cdots,k_\nu)) = (\pi(k_1),\cdots,\pi(k_\nu)), \quad \text{for all } \pi \in \mathrm{Aut}(\mathbb{F}_q)$$
$$\text{and } k_1,\cdots,k_\nu \in \mathbb{F}_q^*;$$

then the semidirect product of $\mathbb{F}_q^* \times \cdots \times \mathbb{F}_q^*$ by $\mathrm{Aut}(\mathbb{F}_q)$ corresponding to φ, denoted by $(\mathbb{F}_q^* \times \cdots \times \mathbb{F}_q^*) \rtimes_\varphi \mathrm{Aut}(\mathbb{F}_q)$, is the group consisting of all elements of the form $(k_1,\cdots,k_\nu,\pi)$, where $k_1,\cdots,k_\nu \in \mathbb{F}_q^*$ and $\pi \in \mathrm{Aut}(\mathbb{F}_q)$, with multiplication defined by

$$(k_1, \cdots, k_\nu, \pi)(k_1', \cdots, k_\nu', \pi') = (k_1\pi(k_1'), \cdots, k_\nu\pi(k_\nu'), \pi\pi').$$

Then the main result about $\mathrm{Aut}(Sp(2\nu, q))$ is as follows.

Theorem 3.4   *Regard $PSp_{2\nu}(\mathbb{F}_q)$ as a subgroup of $\mathrm{Aut}(Sp(2\nu, q))$ and let E be the subgroup of $\mathrm{Aut}(Sp(2\nu, q))$ defined as follows:*

$$E = \{\sigma \in \mathrm{Aut}(Sp(2\nu, q)) : \sigma([e_i]) = [e_i], \sigma([f_i]) = [f_i], i = 1, \cdots, \nu\}.$$

Then:

(1) $\mathrm{Aut}(Sp(2\nu, q)) = PSp_{2\nu}(\mathbb{F}_q) \cdot E$;

(2) *if $\nu = 1$, then E is isomorphic to the symmetric group on $q-1$ elements;*

(3) *if $\nu > 1$, then*

$$E \cong \underbrace{(\mathbb{F}_q^* \times \cdots \times \mathbb{F}_q^*)}_{\nu} \rtimes_\varphi \mathrm{Aut}(\mathbb{F}_q).$$

Proof   (1) Let $\tau \in \mathrm{Aut}(Sp(2\nu, q))$. Suppose that $\tau([e_i]) = [e_i'], \tau([f_i]) = [f_i'], i = 1, \cdots, \nu$. Then $e_i'K^t f_i' \neq 0$, $e_i'K^t e_j' = 0$, $f_i'K^t f_j' = 0$, $i, j = 1, \cdots, \nu$ and $e_i'K^t f_j' = 0, i \neq j, i, j = 1, \cdots, \nu$. We may choose $e_i', f_i', i = 1, \cdots, \nu$, such that $e_i'K^t f_i' = 1, i = 1, \cdots, \nu$. Let

$$A = \begin{pmatrix} e_1 \\ f_1 \\ e_2 \\ f_2 \\ \vdots \\ e_\nu \\ f_\nu \end{pmatrix}, \quad A' = \begin{pmatrix} e_1' \\ f_1' \\ e_2' \\ f_2' \\ \vdots \\ e_\nu' \\ f_\nu' \end{pmatrix}.$$

Then $AK^t A = K = A'K^t A'$. Thus, by [6, Lemma 3.11], there exists $T \in Sp_{2\nu}(\mathbb{F}_q)$ such that $A = A'T$, i.e., $e_i'T = e_i$, $f_i'T = f_i$, $i = 1, \cdots, \nu$. Set $\tau_1 = \sigma_T \tau$. Then $\tau_1([e_i]) = [e_i], \tau_1([f_i]) = [f_i], i = 1, \cdots, \nu$; hence $\tau_1 \in E$. Thus $\tau \in PSp_{2\nu}(\mathbb{F}_q) \cdot E$. It follows that $\mathrm{Aut}(Sp(2\nu, q)) = PSp_{2\nu}(\mathbb{F}_q) \cdot E$.

(2) When $\nu = 1$, it is clear that E is isomorphic to the symmetric group on the $q - 1$ vertices of $Sp(2, q)$ since $Sp(2, q)$ is a complete graph.

(3) Suppose that $\nu > 1$. Our idea for proving the third part of the theorem is to identify some elements of E which form a subgroup of E isomorphic to the semidirect product in the theorem and, then, to show that every element of E has the form of the elements identified before.

Firstly, let us write out some elements of E. Let $k_1, \cdots, k_\nu \in \mathbb{F}_q^*$ and $\pi \in \mathrm{Aut}(\mathbb{F}_q)$. Let $\sigma_{(k_1,\cdots,k_\nu,\pi)}$ be the map which takes any vertex $[a_1, a_2, a_3, a_4, \cdots, a_{2\nu-1}, a_{2\nu}]$ of $Sp(2\nu, q)$ to the vertex

$$[\pi(a_1), k_1\pi(a_2), k_2\pi(a_3), k_1 k_2^{-1}\pi(a_4), \cdots, k_\nu \pi(a_{2\nu-1}), k_1 k_\nu^{-1}\pi(a_{2\nu})].$$

Then it is clear that $\sigma_{(k_1,\cdots,k_\nu,\pi)}$ is well defined. Furthermore, it is easy to see that $\sigma_{(k_1,\cdots,k_\nu,\pi)}$ is injective, but the vertex set of $Sp(2\nu, q)$ is finite; $\sigma_{(k_1,\cdots,k_\nu,\pi)}$ is a bijection from $V(Sp(2\nu, q))$ to itself. Let $\alpha = [a_1, a_2, a_3, a_4, \cdots, a_{2\nu-1}, a_{2\nu}]$, $\beta = [a_1', a_2', a_3', a_4', \cdots, a_{2\nu-1}', a_{2\nu}']$ be two vertices of $Sp(2\nu, q)$. If $\alpha \not\sim \beta$; then, by definition,

$$(a_1 a_2' - a_2 a_1') + (a_3 a_4' - a_4 a_3') + \cdots + (a_{2\nu-1} a_{2\nu}' - a_{2\nu} a_{2\nu-1}') = 0,$$

which implies that

$$(\pi(a_1) k_1 \pi(a_2') - \pi(a_2) k_1 \pi(a_1')) + (k_2 \pi(a_3) k_1 k_2^{-1} \pi(a_4') - k_1 k_2^{-1} \pi(a_4) k_2 \pi(a_3'))$$
$$+ \cdots + (k_\nu \pi(a_{2\nu-1}) k_1 k_\nu^{-1} \pi(a_{2\nu}') - k_1 k_\nu^{-1} \pi(a_{2\nu}) k_\nu \pi(a_{2\nu-1}')) = 0,$$

i.e., $\sigma_{(k_1,\cdots,k_\nu,\pi)}(\alpha) \not\sim \sigma_{(k_1,\cdots,k_\nu,\pi)}(\beta)$. Since the edges set of $Sp(2\nu, q)$ is finite, $\alpha \not\sim \beta$ if and only if $\sigma_{(k_1,\cdots,k_\nu,\pi)}(\alpha) \not\sim \sigma_{(k_1,\cdots,k_\nu,\pi)}(\beta)$. Hence $\sigma_{(k_1,\cdots,k_\nu,\pi)} \in \mathrm{Aut}(Sp(2\nu, q))$. Note that $\sigma_{(k_1,\cdots,k_\nu,\pi)}([e_i]) = [e_i]$, $\sigma_{(k_1,\cdots,k_\nu,\pi)}([f_i]) = [f_i]$, $i = 1, \cdots, \nu$, hence, $\sigma_{(k_1,\cdots,k_\nu,\pi)} \in E$.

If we define a map h as $(k_1, \cdots, k_\nu, \pi) \mapsto \sigma_{(k_1,\cdots,k_\nu,\pi)}$, then it is easy to verify that h is a group homomorphism from $(\mathbb{F}_q^* \times \cdots \times \mathbb{F}_q^*) \times_\varphi \mathrm{Aut}(\mathbb{F}_q)$ to E. It is also easy to see that if $(k_1, \cdots, k_\nu, \pi) \neq (k_1', \cdots, k_\nu', \pi')$ then $\sigma_{(k_1,\cdots,k_\nu,\pi)} \neq \sigma_{(k_1',\cdots,k_\nu',\pi')}$. In order to prove the third part of the theorem, we will show that h is a group isomorphism. It remains to show that every element of E is of the form $\sigma_{(k_1,\cdots,k_\nu,\pi)}$.

Suppose that $\sigma \in E$. Note that if $\sigma([a_1, a_2, \cdots, a_{2\nu}]) = [b_1, b_2, \cdots, b_{2\nu}]$, then $a_{2i-1} \neq 0$ if and only if $[a_1, a_2, \cdots, a_{2\nu}] \sim [f_i]$ and $a_{2i} \neq 0$ if and only if $[a_1, a_2, \cdots, a_{2\nu}] \sim [e_i]$, and similar results are also true for b_i. But $\sigma([e_i]) = [e_i]$ and $\sigma([f_i]) = [f_i]$, it follows that $a_i = 0$ if and only if $b_i = 0$. For any vertex $[a_1, a_2, \cdots, a_{2\nu}]$, if $a_1 = \cdots = a_{i-1} = 0$ and $a_i \neq 0$ then $[a_1, a_2, \cdots, a_{2\nu}]$ can be uniquely written as $[0, \cdots, 0, 1, a'_{i+1}, \cdots, a'_{2\nu}]$ and $\sigma([a_1, a_2, \cdots, a_{2\nu}])$ can be uniquely written as $[0, \cdots, 0, 1, b'_{i+1}, \cdots, b'_{2\nu}]$. Let us show how to determine $b'_{i+1}, \cdots, b'_{2\nu}$ from $a'_{i+1}, \cdots, a'_{2\nu}$. We will use frequently the fact that, for any vertices $[\alpha], [\beta]$, if $[\alpha] \not\sim [\beta]$ then $\sigma([\alpha]) \not\sim \sigma([\beta])$.

In the following, we will denote $[a_1, a'_1, a_2, a'_2, \cdots, a_\nu, a'_\nu]$ by $\sum_{i=1}^{\nu} a_i[e_i] + \sum_{i=1}^{\nu} a'_i[f_i]$; for example, $[a, b, 0, \cdots, 0]$ is denoted by $a[e_1] + b[f_1]$. Since σ is a bijection from $V(Sp(2\nu, q))$ to itself, we have permutations $\pi_i, i = 2, \cdots, 2\nu$, of $\mathbb{F}_q$ with $\pi(0) = 0$ such that

$$\sigma([e_1] + a_{2i-1}[e_i]) = [e_1] + \pi_{2i-1}(a_{2i-1})[e_i],$$
$$\sigma([e_1] + a_{2i}[f_i]) = [e_1] + \pi_{2i}(a_{2i})[f_i].$$

We firstly consider the cases $\sigma([0, 1, a_3, \cdots, a_{2\nu}])$ and $\sigma([1, a_2, a_3, \cdots, a_{2\nu}])$. Let $\sigma([0, 1, a_3, \cdots, a_{2\nu}]) = [0, 1, a'_3, \cdots, a'_{2\nu}]$ and $j \geq 1$. If $a_{2j+1} \neq 0$; then, from $[0, 1, a_3, \cdots, a_{2\nu}] \not\sim [e_1] + a_{2j+1}^{-1}[f_{j+1}]$ we have $[0, 1, a'_3, \cdots, a'_{2\nu}] \not\sim [e_1] + \pi_{2j+2}(a_{2j+1}^{-1})[f_{j+1}]$; hence, $a'_{2j+1} = \pi_{2j+2}(a_{2j+1}^{-1})^{-1}$. If $a_{2j+2} \neq 0$, then from $[0, 1, a_3, \cdots, a_{2\nu}] \not\sim [e_1] - a_{2j+2}^{-1}[e_{j+1}]$ we have $[0, 1, a'_3, \cdots, a'_{2\nu}] \not\sim [e_1] + \pi_{2j+1}(-a_{2j+2}^{-1})[e_{j+1}]$, hence, $a'_{2j+2} = -\pi_{2j+1}(-a_{2j+2}^{-1})^{-1}$. Thus

$$\sigma([0, 1, a_3, \cdots, a_{2\nu}]) = [0, 1, a'_3, \cdots, a'_{2\nu}], \tag{1}$$

where $a'_{2j+1} = \pi_{2j+2}(a_{2j+1}^{-1})^{-1}$ if $a_{2j+1} \neq 0$ and $a'_{2j+2} = -\pi_{2j+1}(-a_{2j+2}^{-1})^{-1}$ if $a_{2j+2} \neq 0$.

For the case $\sigma([1, a_2, a_3, \cdots, a_{2\nu}])$, let $\sigma([1, a_2, a_3, \cdots, a_{2\nu}]) = [1, a''_2, a''_3, \cdots, a''_{2\nu}]$. From $[1, a_2, a_3, \cdots, a_{2\nu}] \not\sim [e_1] + a_2[f_1]$ we get $[1, a''_2, a''_3, \cdots, a''_{2\nu}] \not\sim [e_1] + \pi_2(a_2)[f_1]$; hence, $a''_2 = \pi_2(a_2)$. Let $j \geq 1$. If $a_{2j+1} \neq 0$, then, from $[1, a_2, a_3, \cdots, a_{2\nu}] \not\sim [f_1] - a_{2j+1}^{-1}[f_{j+1}]$ and $\sigma([f_1] - a_{2j+1}^{-1}[f_{j+1}]) = [f_1] -$

$\pi_{2j+1}(a_{2j+1})^{-1}[f_{j+1}]$ as shown above, we have $[1, a_2'', a_3'', \cdots, a_{2\nu}''] \not\sim [f_1] - \pi_{2j+1}(a_{2j+1})^{-1}[f_{j+1}]$; hence, $a_{2j+1}'' = \pi_{2j+1}(a_{2j+1})$. Similarly, if $a_{2j+2} \neq 0$, then from $[1, a_2, a_3, \cdots, a_{2\nu}] \not\sim [f_1] + a_{2j+2}^{-1}[e_{j+1}]$ we have $[1, a_2'', a_3'', \cdots, a_{2\nu}''] \not\sim [f_1] + \pi_{2j+2}(a_{2j+2})^{-1}[e_{j+1}]$; hence, $a_{2j+2}'' = \pi_{2j+2}(a_{2j+2})$. Thus, for any $a_2, a_3, \cdots, a_{2\nu} \in \mathbb{F}_q$,

$$\sigma([1, a_2, a_3, \cdots, a_{2\nu}]) = [1, \pi_2(a_2), \pi_3(a_3), \cdots, \pi_{2\nu}(a_{2\nu})]. \tag{2}$$

Then, let $i \geqslant 2$; we discuss the general cases $\sigma([0, \cdots, 0, 1, a_{2i+1}, \cdots, a_{2\nu}])$ and $\sigma([0, \cdots, 0, 1, a_{2i}, \cdots, a_{2\nu}])$. The above results of case $i = 1$ will be used. Let $\sigma([0, \cdots, 0, 1, a_{2i+1}, \cdots, a_{2\nu}]) = [0, \cdots, 0, 1, a_{2i+1}', \cdots, a_{2\nu}']$ and $j \geqslant i$. If $a_{2j+1} \neq 0$, then, from

$$[0, \cdots, 0, 1, a_{2i+1}, \cdots, a_{2\nu}] \not\sim [e_1] + [e_i] + a_{2j+1}^{-1}[f_{j+1}]$$

and $\sigma([e_1] + [e_i] + a_{2j+1}^{-1}[f_{j+1}]) = [e_1] + \pi_{2i-1}(1)[e_i] + \pi_{2j+2}(a_{2j+1}^{-1})[f_{j+1}]$ as shown above, we have

$$[0, \cdots, 0, 1, a_{2i+1}', \cdots, a_{2\nu}'] \not\sim [e_1] + \pi_{2i-1}(1)[e_i] + \pi_{2j+2}(a_{2j+1}^{-1})[f_{j+1}];$$

hence, $a_{2j+1}' = \pi_{2i-1}(1)\pi_{2j+2}(a_{2j+1}^{-1})^{-1}$. Similarly, if $a_{2j+2} \neq 0$, then from

$$[0, \cdots, 0, 1, a_{2i+1}, \cdots, a_{2\nu}] \not\sim [e_1] + [e_i] - a_{2j+2}^{-1}[e_{j+1}]$$

we have

$$[0, \cdots, 0, 1, a_{2i+1}', \cdots, a_{2\nu}'] \not\sim [e_1] + \pi_{2i-1}(1)[e_i] + \pi_{2j+1}(-a_{2j+2}^{-1})[e_{j+1}];$$

hence, $a_{2j+2}' = -\pi_{2i-1}(1)\pi_{2j+1}(-a_{2j+2}^{-1})^{-1}$. Thus,

$$\sigma([0, \cdots, 0, 1, a_{2i+1}, \cdots, a_{2\nu}]) = [0, \cdots, 0, 1, a_{2i+1}', \cdots, a_{2\nu}'], \tag{3}$$

where $a_{2j+1}' = \pi_{2i-1}(1)\pi_{2j+2}(a_{2j+1}^{-1})^{-1}$ if $a_{2j+1} \neq 0$ and $a_{2j+2}' = -\pi_{2i-1}(1) \cdot \pi_{2j+1}(-a_{2j+2}^{-1})^{-1}$ if $a_{2j+2} \neq 0$.

Finally, for the case $\sigma([0, \cdots, 0, 1, a_{2i}, \cdots, a_{2\nu}])$, let $\sigma([0, \cdots, 0, 1, a_{2i}, \cdots, a_{2\nu}]) = [0, \cdots, 0, 1, a_{2i}'', \cdots, a_{2\nu}'']$. From

$$[0, \cdots, 0, 1, a_{2i}, \cdots, a_{2\nu}] \not\sim [e_1] + [e_i] + a_{2i}[f_i]$$

we get
$$[0,\cdots,0,1,a''_{2i},\cdots,a''_{2\nu}] \not\sim [e_1] + \pi_{2i-1}(1)[e_i] + \pi_{2i}(a_{2i})[f_i];$$

hence, $a''_{2i} = \pi_{2i-1}(1)^{-1}\pi_{2i}(a_{2i})$. Let $j \geqslant i$. If $a_{2j+1} \neq 0$, then from
$$[0,\cdots,0,1,a_{2i},\cdots,a_{2\nu}] \not\sim [f_i] - a_{2j+1}^{-1}[f_{j+1}]$$

we have
$$[0,\cdots,0,1,a''_{2i},\cdots,a''_{2\nu}] \not\sim [f_i] - \pi_{2i-1}(1)^{-1}\pi_{2j+1}(a_{2j+1})^{-1}[f_{j+1}];$$

hence, $a''_{2j+1} = \pi_{2i-1}(1)^{-1}\pi_{2j+1}(a_{2j+1})$. If $a_{2j+2} \neq 0$, then from
$$[0,\cdots,0,1,a_{2i},\cdots,a_{2\nu}] \not\sim [f_i] + a_{2j+2}^{-1}[e_{j+1}]$$

we have
$$[0,\cdots,0,1,a''_{2i},\cdots,a''_{2\nu}] \not\sim [f_i] + \pi_{2i-1}(1)\pi_{2j+2}(a_{2j+2})^{-1}[e_{j+1}];$$

hence $a''_{2j+2} = \pi_{2i-1}(1)^{-1}\pi_{2j+2}(a_{2j+2})$. Thus, for any $a_{2i}, a_{2i+1},\cdots,a_{2\nu} \in \mathbb{F}_q$,
$$\begin{aligned}&\sigma([0,\cdots,0,1,a_{2i},a_{2i+1},\cdots,a_{2\nu}])\\&=[0,\cdots,0,1,\pi_{2i-1}(1)^{-1}\pi_{2i}(a_{2i}),\pi_{2i-1}(1)^{-1}\pi_{2i+1}(a_{2i+1}),\cdots,\\&\quad\pi_{2i-1}(1)^{-1}\pi_{2\nu}(a_{2\nu})].\end{aligned} \tag{4}$$

Having represented σ by $\pi_i, i = 2,\cdots,2\nu$, let us discuss some properties of π_i.

Lemma 3.5 (1) For any $i \geqslant 1$ and $a \in \mathbb{F}_q$,
$$\pi_{2i+1}(1)\pi_{2i+2}(a) = \pi_{2i+2}(1)\pi_{2i+1}(a) = \pi_2(a).$$

(2) For any $i \geqslant 2$ and $a,b \in \mathbb{F}_q$,
$$\begin{aligned}\pi_i(a+b) &= \pi_i(a) + \pi_i(b);\\ \pi_i(-a) &= -\pi_i(a);\end{aligned}$$

$$\pi_i(ab) = \pi_i(a)\pi_i(b)\pi_i(1)^{-1};$$
$$\pi_i(a^{-1}) = \pi_i(a)^{-1}\pi_i(1)^2 \text{ if } a \neq 0.$$

Proof (1) We may assume that $a \neq 0$. Since $[e_1] + a[e_{i+1}] + a[f_{i+1}] \not\sim [e_{i+1}] + [f_{i+1}]$, it follows that $\sigma([e_1]+a[e_{i+1}]+a[f_{i+1}]) \not\sim \sigma([e_{i+1}]+[f_{i+1}])$, but

$$\sigma([e_1] + a[e_{i+1}] + a[f_{i+1}]) = [e_1] + \pi_{2i+1}(a)[e_{i+1}] + \pi_{2i+2}(a)[f_{i+1}],$$
$$\sigma([e_{i+1}] + [f_{i+1}]) = [e_{i+1}] + \pi_{2i+1}(1)^{-1}\pi_{2i+2}(1)[f_{i+1}];$$

we have that

$$\pi_{2i+1}(1)^{-1}\pi_{2i+2}(1)\pi_{2i+1}(a) - \pi_{2i+2}(a) = 0,$$

i.e.,

$$\pi_{2i+1}(1)\pi_{2i+2}(a) = \pi_{2i+2}(1)\pi_{2i+1}(a).$$

Similarly, since $[e_1]+a[f_1]+[e_{i+1}] \not\sim [e_1]+a[f_{i+1}]$, we have that $[e_1]+\pi_2(a)[f_1]+\pi_{2i+1}(1)[e_{i+1}] \not\sim [e_1]+\pi_{2i+2}(a)[f_{i+1}]$, hence, $\pi_{2i+1}(1)\pi_{2i+2}(a) = \pi_2(a)$.

(2) From $[e_1] + (a+b)[f_1] + [e_2] \not\sim [e_1] + a[f_1] + b[f_2]$ we have that

$$[e_1] + \pi_2(a+b)[f_1] + \pi_3(1)[e_2] \not\sim [e_1] + \pi_2(a)[f_1] + \pi_4(b)[f_2].$$

Then $\pi_2(a) - \pi_2(a+b) + \pi_3(1)\pi_4(b) = 0$, but $\pi_3(1)\pi_4(b) = \pi_2(b)$; hence, $\pi_2(a+b) = \pi_2(a) + \pi_2(b)$. It turns out from (1) that this equality holds for all $i \geqslant 2$. Thus $\pi_i(-a) = -\pi_i(a)$ as $\pi_i(0) = 0$.

For multiplication, let $i \geqslant 1$; from $[e_1]+b[e_{i+1}]+ab[f_{i+1}] \not\sim [e_{i+1}]+a[f_{i+1}]$ we get that

$$[e_1] + \pi_{2i+1}(b)[e_{i+1}] + \pi_{2i+2}(ab)[f_{i+1}] \not\sim [e_{i+1}] + \pi_{2i+1}(1)^{-1}\pi_{2i+2}(a)[f_{i+1}];$$

hence, $\pi_{2i+1}(b)\pi_{2i+1}(1)^{-1}\pi_{2i+2}(a) - \pi_{2i+2}(ab) = 0$, but $\pi_{2i+1}(b)\pi_{2i+1}(1)^{-1} = \pi_{2i+2}(b)\pi_{2i+2}(1)^{-1}$. Thus

$$\pi_{2i+2}(ab) = \pi_{2i+2}(a)\pi_{2i+2}(b)\pi_{2i+2}(1)^{-1}.$$

It follows from $\pi_{2i+1}(1)\pi_{2i+2}(a) = \pi_{2i+2}(1)\pi_{2i+1}(a)$ and $\pi_{2i+1}(1)\pi_{2i+2}(1) = \pi_{2i+2}(1)\pi_{2i+1}(1)$ that the above equality also holds for $2i+1$. It remains to consider π_2.

We have

$$\begin{aligned}\pi_2(ab) &= \pi_3(1)\pi_4(ab) \\ &= \pi_3(1)\pi_4(1)^{-1}\pi_4(a)\pi_4(b) \\ &= \pi_3(1)^{-1}\pi_4(1)^{-1}\pi_2(a)\pi_2(b) \\ &= \pi_2(a)\pi_2(b)\pi_2(1)^{-1}.\end{aligned}$$

Finally, if $a \neq 0$, then from $\pi_i(1) = \pi_i(aa^{-1}) = \pi_i(a)\pi_i(a^{-1})\pi_i(1)^{-1}$ we obtain that $\pi_i(a^{-1}) = \pi_i(a)^{-1}\pi_i(1)^2$; then the proof of the lemma is complete. $\square$

We continue the proof of the theorem. Let us denote the identity automorphism on $\mathbb{F}_q$ by π_1. Then when $i = 1$, (3) reduces to (1) and (4) reduces to (2). Therefore (3) and (4) hold for all i, where $1 \leqslant i \leqslant \nu$. By the above lemma, for any $i \geqslant 1$, we can rewrite (3) in the form of (4) as follows. In (3), for any $j \geqslant i$, we have

$$\begin{aligned}a'_{2j+1} &= \pi_{2i-1}(1)\pi_{2j+2}(a_{2j+1}^{-1})^{-1} \\ &= \pi_{2i-1}(1)\pi_{2j+2}(a_{2j+1})\pi_{2j+2}(1)^{-2} \\ &= \pi_{2i-1}(1)\pi_{2j+1}(1)^{-1}\pi_{2j+2}(1)^{-1}\pi_{2j+1}(a_{2j+1}) \\ &= \pi_{2i-1}(1)\pi_2(1)^{-1}\pi_{2j+1}(a_{2j+1}) \\ &= \pi_{2i}(1)^{-1}\pi_{2j+1}(a_{2j+1})\end{aligned}$$

and

$$\begin{aligned}a'_{2j+2} &= -\pi_{2i-1}(1)\pi_{2j+1}(-a_{2j+2}^{-1})^{-1} \\ &= \pi_{2i-1}(1)\pi_{2j+1}(a_{2j+2}^{-1})^{-1} \\ &= \pi_{2i-1}(1)\pi_{2j+1}(a_{2j+2})\pi_{2j+1}(1)^{-2} \\ &= \pi_{2i-1}(1)\pi_{2j+1}(1)^{-1}\pi_{2j+2}(1)^{-1}\pi_{2j+2}(a_{2j+2})\end{aligned}$$

$$=\pi_{2i-1}(1)\pi_2(1)^{-1}\pi_{2j+2}(a_{2j+2})$$
$$=\pi_{2i}(1)^{-1}\pi_{2j+2}(a_{2j+2}).$$

Hence, for any $a_{2i+1}, \cdots, a_{2\nu} \in \mathbb{F}_q$,

$$\sigma([0, \cdots, 0, 1, a_{2i+1}, \cdots, a_{2\nu}])$$
$$=[0, \cdots, 0, 1, \pi_{2i}(1)^{-1}\pi_{2i+1}(a_{2i+1}), \cdots, \pi_{2i}(1)^{-1}\pi_{2\nu}(a_{2\nu})], \quad (5)$$

which is of the same form as (4).

Now let $k_1 = \pi_2(1)$, $\pi = k_1^{-1}\pi_2$, $k_2 = \pi_3(1)$, $k_3 = \pi_5(1), \cdots, k_\nu = \pi_{2\nu-1}(1)$. Then $\pi \in \text{Aut}(\mathbb{F}_q)$, $\pi_2 = k_1\pi$, $\pi_3 = k_2\pi$, $\pi_4 = k_1 k_2^{-1}\pi, \cdots, \pi_{2\nu-1} = k_\nu\pi$, $\pi_{2\nu} = k_1 k_\nu^{-1}\pi$. Assembling (4) and (5), we obtain

$$\sigma([a_1, a_2, a_3, a_4, \cdots, a_{2\nu-1}, a_{2\nu}])$$
$$=[\pi(a_1), k_1\pi(a_2), k_2\pi(a_3), k_1 k_2^{-1}\pi(a_4), \cdots, k_\nu\pi(a_{2\nu-1}), k_1 k_\nu^{-1}\pi(a_{2\nu})].$$

Hence $\sigma = h(k_1, \cdots, k_\nu, \pi)$ as required; the proof of the theorem is complete. □

Corollary 3.6 When $\nu = 1$,

$$|\text{Aut}(Sp(2, q))| = q(q^2 - 1) \cdot (q - 2)!,$$

and when $\nu \geq 2$,

$$|\text{Aut}(Sp(2\nu, q))| = q^{\nu^2} \prod_{i=1}^{\nu}(q^{2i} - 1) \cdot [\mathbb{F}_q : \mathbb{F}_p].$$

Proof Note that $PSp_{2\nu}(\mathbb{F}_q) \cap E$ consists of σ which is reduced from some matrix of the form $\text{diag}(k_1, l_1, k_2, l_2, \cdots, k_\nu, l_\nu)$, with $k_i l_i = 1, i = 1, \cdots, \nu$. Thus $|PSp_{2\nu}(\mathbb{F}_q) \cap E| = \frac{1}{2}(q-1)^\nu$. Hence

$$|\text{Aut}(Sp(2\nu, q))| = \frac{|PSp_{2\nu}(\mathbb{F}_q)||E|}{|PSp_{2\nu}(\mathbb{F}_q) \cap E|}$$

$$= \frac{\frac{1}{2}q^{\nu^2} \prod_{i=1}^{\nu}(q^{2i} - 1) \cdot |E|}{\frac{1}{2}(q-1)^\nu}.$$

Thus, when $\nu = 1$, $|\text{Aut}(Sp(2,q))| = q(q^2-1) \cdot (q-2)!$, and when $\nu \geq 2$,

$$|\text{Aut}(Sp(2\nu,q))|$$
$$= \frac{\frac{1}{2}q^{\nu^2}\prod_{i=1}^{\nu}(q^{2i}-1) \cdot (q-1)^{\nu} \cdot |\text{Aut}(\mathbb{F}_q)|}{\frac{1}{2}(q-1)^{\nu}}$$
$$= q^{\nu^2}\prod_{i=1}^{\nu}(q^{2i}-1) \cdot |\text{Aut}(\mathbb{F}_q)|$$
$$= q^{\nu^2}\prod_{i=1}^{\nu}(q^{2i}-1) \cdot [\mathbb{F}_q : \mathbb{F}_q],$$

as it is well known that $|\text{Aut}(\mathbb{F}_q)| = [\mathbb{F}_q : \mathbb{F}_p]$ where $p = \text{char}(\mathbb{F}_q)$. □

Acknowledgement

Both authors are supported by the National Natural Science Foundation of China.

References

[1] R. H. Dye, Partitions and their stabilizers for line complexes and quadrics, Ann. Mat. Pura Appl. 114 (1977) 173–194.

[2] C. Godsil, G. Royle, Chromatic number and the 2-rank of a graph, J. Combin. Theory Ser. B 81 (2001) 142–149.

[3] C. Godsil, G. Royle, Algebraic Graph Theory, Graduate Texts in Mathematics, vol. 207, Springer-Verlag, 2001.

[4] J.J. Rotman, Projective planes, graphs, and simple algebras, J. Algebra 155 (1993) 267–289.

[5] J.J. Rotman, P.M. Weichsel, Simple Lie algebras and graphs, J. Algebra 169 (1994) 775–790.

[6] Z. Wan, Geometry of Classical Groups over Finite Fields, 2nd edition, Science Press, Beijing, New York, 2002.

《典型群》序[①]

早在 1949 年,本书作者之一就有了写这样一本书的轮廓,希望根据这个轮廓组织一个讨论班,和一批大学四年级及刚毕业的同学在一起,使他们边学习边搞研究,集体地较整套地来进行这一领域的研究工作,一来可以使他们在工作过程中逐步地扩充自己的知识领域,另一方面可以让他们习作一些研究,预计在计划完成之后,可以给典型群论、射影几何学、矩阵论及群表示论等数学分支以一个不同的面貌. 1950 年年初,当他在北京清华大学执教时,组织了这样一个讨论班. 讨论班进行到 1951 年暑假,在讨论班里他完成了本书前六章的初稿. 接着,在 1951 年下半年和 1957 年上半年,他又在中国科学院数学研究所代数讨论班里两度报告了本书前六章的大部分章节,并进行了一些修改. 随后,从 1959 年下半年起,本书后一作者又在数学研究所代数讨论班里报告了前六章的部分章节,并根据他所体会的前六章的精神与方法续写了本书的后六章. 这就是本书简单的写作经过.

简要地可以这样说,体上的矩阵是一个值得注意的对象,因为它是一个不太失去普遍性的抽象事物,但同时又和成果丰富的具体的域上的矩阵论距离不远. 当然,结合环、李环和柔丹环中有趣的部分又都有矩阵形式,而线性群、正交群、辛群、洛伦兹群也都是矩阵群;几何学中线几何、圆几何、格拉斯曼几何都有矩阵的表示法. 更如多复变函数论的典型域也离不开矩阵的表达形式. 这些形成了我们的工作背景.

仅仅找到一个值得研究的对象,而没有处理的方法那也还是空话. 本书中提供了一些方法,这些方法是初步的,有待改进、补充和发展,只有在发展过程中才能把方法搞得更完备.

1950 年本书作者之一选择这个主题的原因之一是为了易于训练干部. 预

[①] 华罗庚、万哲先合著的《典型群》一书 1963 年由上海科学技术出版社出版,2010 年由科学出版社以《华罗庚文集》代数卷 I 的形式再版.

备知识需要得少，可以从简单处着手，从具体处着手；发展前途不太小，通过这一系列的研究也可以熟悉代数学、几何学中不少分支，可以从宽广处着眼，从抽象处着眼. 换言之，开始时不受基础的限制，终了时不致局促于太仄狭的领域之中.

匆匆已经十年，这计划还只能说在第一阶段中完成了第一部分而已. 更重要的工作还有待于今后的努力. 这决不是一个完整的东西，而仅仅是一个开始. 这是一个阶梯中的一级，读者必须想想前面几级 —— 实数域、复数域、有限域及四元数体上的情况，读者更必须看看后面几级 —— 体上矩阵的环和群的构造、不用连续性的群表示论等等，这样才不至于为本书引入歧途. 长期局限于本书范围内的工作将不是作者的本意，但我们认为搞清这些对象和方法对学好典型群论、射影几何学等都能有所帮助.

我们感谢王仰贤、应玫茜、徐诚浩等同志，在本书付印之前，他们分头阅读了本书手稿的各部分，进行了核算，并提出了一系列宝贵的修改意见.

华罗庚　万哲先

1962 年 8 月于北京

《李代数》序[①]

初 版 序

　　1961年秋至1963年春作者在中国科学院数学研究所李群讨论班上陆续作了一些专题报告，本书就是根据这些报告的讲稿改编而成. 内容包括复半单李代数的经典理论，即它的结构、自同构、表示和实形. 当时, 作者的目的是和参加讨论班的同志们共同学习李代数的基础知识, 为进一步学习李群及李代数的近代文献打下基础. 这些专题报告, 主要参考邓金的《半单纯李氏代数的结构》(曾肯成译, 北京: 科学出版社, 1954) 和 Seminaire Sophus Lie 的讲义 *Théorie de Algèbres de Lie et Topologie de Groupes de Lie* (Paris, 1955). 邓金的书叙述清楚, 易于被初学者领会, 遗憾的是内容太少; Seminaire Sophus Lie 的讲义内容较丰富, 但是要求读者具有较多的预备知识. 两书虽各具优点, 但都不能完全满足我国初学者的需要, 于是, 作者就想到要写一本适应这种需要的书. 这就是这本书的来历.

　　李代数是 S. Lie 作为研究后来以他命名的李群的代数工具而引进的. 在李代数经典理论方面有重要贡献者, 除 S. Lie 本人外, 当推 W. Killing, E. Cartan 和 H. Weyl 等人. 虽然本书为了初学者的方便, 在叙述上尽可能少地涉及李群, 却应该指出, 李代数经典理论的重要性主要在于它对李群的应用. 另一方面, 本书的大部分内容, 都已推广到特征 0 的代数封闭域上的李代数, 而且一部分结果也已推广到特征 0 的任意域上的李代数. 但我们在本书中却仅对复数域上的李代数来叙述, 这是因为复数域上的李代数理论是最基本的, 同时只要求读者具备线性代数知识就能阅读本书绝大部分内容也是一个限制.

　　[①] 万哲先《李代数》一书 1964 年由科学出版社出版, 1972 年商务印书馆香港分馆又再次出版. 1975 年本书英译本由英国 Pergamon 出版社出版. 1978 年科学出版社又发行了第二次印刷本. 2013 年由高等教育出版社出版了第二版.

作者感谢参加李群讨论班的同志们. 当作者报告时, 他们的意见以及和他们共同讨论使作者学到很多东西, 也使本书的某些叙述得以改进并消除了许多错误. 作者特别感谢李根道同志, 他还帮助作者校对本书.

万哲先

1963 年 4 月

第 二 版 序

　　本书 1964 年由科学出版社出版, 1972 年商务印书馆香港分馆又再次出版. 1975 年本书英译本由英国 Pergamon 出版社出版. 1978 年科学出版社又发行了第二次印刷本.

　　现在高等教育出版社出版了第二版. 在这一版中, 作者对本书的体例格式进行了便于查询的修改, 改正了第一版某些排版错误, 并修改了部分定理的证明, 使得本书结构更清晰, 更具可读性. 在出版过程中, 高等教育出版社的王丽萍、李鹏等同志给我热情、耐心与细致的帮助. 我谨对他们表示衷心的感谢.

万哲先

2013 年 5 月

《有限几何与不完全区组设计的一些研究》序[①]

本书包括我们于 1963 年至 1964 年间在有限几何和利用有限几何来构作部分平衡不完全区组设计两方面得到的一些结果,其中一部分结果尚未发表过.

应用有限几何来构作不完全区组设计已有二十余年的历史. 北京大学的同志们于 1963 年 3 月在北京数学会代数组的学术报告会上作了 "试验设计中的代数问题" 的专题报告. 在报告中,他们介绍了应用有限几何来构作不完全区组设计这一课题的发展情况,从而引起本书前一作者的注意. 当本书前一作者在这方面进行了一些初步探索之后,其余三位作者乘在中国科学技术大学进行毕业论文的机会,也参加到这项工作中来.

本书共分十章,前五章讨论有限几何的计数定理. 在第一章里,我们介绍了经典的计数定理,即在 q 个元素的有限域 F_q 上的 n 维向量空间 $V_n(F_q)$ 中, m 维 $(1 \leqslant m \leqslant n)$ 子空间的个数是

$$\frac{(q^n-1)(q^n-q)\cdots(q^n-q^{m-1})}{(q^m-1)(q^m-q)\cdots(q^m-q^{m-1})}$$

这一结果. 大家都知道, $V_n(F_q)$ 中 m 维子空间的全体在一般线性群 $GL_n(F_q)$ 作用下组成一个可迁集. $GL_n(F_q)$ 是有限域上的一类典型群,因此,自然要研究 $V_n(F_q)$ 中子空间在有限域上其他类型的典型群 (即辛群、酉群和正交群) 作用下分成的可迁集中子空间的个数. 这就是本书第二、三、四、五章的主要内容. 在这几章里,我们计算了有限域上辛几何、酉几何和正交几何的各个可迁集中子空间的个数,也计算了这几种几何中一个给定的子空间所包含的另一个可迁集中子空间的个数. 我们把这些结果分别叫做有限域上辛几何、酉几何和正交几何的计数定理.

[①] 万哲先、戴宗铎、冯绪宁、阳本傅合著的《有限几何与不完全区组设计的一些研究》一书 1966 年由科学出版社出版.

在本书后五章里，我们利用有限域上上述各种类型的几何而构作了一些结合方案和部分平衡不完全区组设计，并计算了它们的参数，其中一部分参数的计算依赖于前五章得到的计数定理. 这几章的一部分结果，北京大学的同志们曾用另外的方法独立得到. 在这几章里，我们主要是选取有限域上辛几何、酉几何和正交几何中 1 维迷向 (或奇异) 子空间作为处理，以及选取极大全迷向 (或全奇异) 子空间作为处理来构作结合方案和部分平衡不完全区组设计，并计算了它们的参数. 自然，选取这几种几何中其他类型子空间的可迁集作为处理也可以构作结合方案和部分平衡不完全区组设计，这样得到的结合方案和部分平衡不完全区组设计的参数的计算刚刚开始，看来还有很多工作可以进行.

在整个工作进行过程中，我们得到中国科学院数学研究所和中国科学技术大学的大力支持，特此致谢.

北京大学的同志们曾和我们进行多次有益的讨论，特此表示感谢.

我们特别希望得到读者的批评和指正.

万哲先　戴宗铎　冯绪宁　阳本傅
1965 年 7 月

《代数和编码》序[①]

第 一 版 序

代数学是数学的重要基础分支, 有着上千年的悠久历史. 近百年来, 它的发展异常迅速, 积累了丰富的内容, 对数学的近代发展有显著的影响, 同时在数学的其他分支及自然科学的许多部门里都有着重要的应用. 但它对工程技术的应用则是比较间接的. 最近二十年来, 这种情况却有了显著的变化, 代数学的一些成果被成功地直接应用到一些新的工程技术领域中去, 而这些应用也给代数学带来了新的研究课题. 四十年代末期出现了布尔 (Boole) 代数对开关电路的分析与综合以及对电子计算机逻辑设计的应用. 五十年代则出现了抽象代数, 特别是有限域的理论在编码理论中的应用. 此外, 代数学也是由于近代新技术的需要而诞生的时序电路和自动机理论、系统理论以及程序语言和数理语言学等新学科里的重要工具.

虽然人类在通信中采用种种明码和密码的历史相当久远, 但编码理论可以认为是电子技术飞速发展以后, 针对当代数字通信和数字存储等的具体需要, 于五十年代发展成为一门面目全新的应用数学, 目前已有丰富的内容.

本书主要介绍编码理论所用的代数基础知识, 并在这一基础上阐述了编码理论中的某些课题, 以利于有关工程技术人员与数学工作者掌握编码理论. 全书共分五章, 前两章介绍了编码理论所用到的最起码的一些代数基础知识, 包括抽象代数, 特别是有限域方面的知识和线性代数方面的知识; 次两章分别就编码理论中的两类码, 即伪随机序列和纠错码作了导引性的介绍, 重点在于阐明基本概念; 最后一章介绍了编码理论中出现的几个代数问题. 另外, 为不熟悉集合和映射这两个数学基本概念以及整数的分解的读者编写了两个附录附在书末, 在前一附录里也介绍了本书采用的有关集合和映射的一些

[①] 万哲先著《代数和编码》一书 1976 年由科学出版社出版, 1986 年又出版了修订本 (即第二版), 并多次印刷. 2007 年由高等教育出版社出版了第三版.

符号.

  本书初稿是作者于 1973 年初在中国科学院数学研究所为一些单位的工程技术科研人员举办的代数和编码学习班上的讲稿. 参加学习班的各单位的同志对本书初稿提了许多宝贵意见, 作者根据这些意见对初稿进行了修改和补充, 作者谨向这些同志表示感谢.

  编码理论的历史虽然很短, 但内容却非常丰富. 作者是编码理论的一个初学者, 按本书的体系来介绍编码理论所用到的代数知识还是一个尝试, 作者抱着虚心向国内有关同志学习的愿望, 怀着抛砖引玉的心情, 编写了这本书, 希望得到读者的批评指正, 也希望今后国内有更好的、更适宜于我国读者需要的这方面的书籍出版.

万哲先

1973 年 4 月 2 日

修订版前言

  在第二章里, 我们增加了多项式矩阵和矩阵的相似两节, 包括了矩阵的初等因子的理论和矩阵的有理标准形. 这是因为在第三章里, 我们增加了自律线性时序线路一节, 在自律线性时序线路的讨论中要用到它们. 在第三章里我们还增加了 q 元周期序列的几种表示法, 即形式幂级数表示法、有理分式表示法和根表示法, 因为它们是讨论 q 元周期序列的几种很方便的工具. 另外, 我们还做了一些其他补充.

  在本书第一版出版后, 曾收到许多读者的来信, 作者根据这些来信, 作了一些小修改, 在这里谨向所有来信的同志表示感谢.

万哲先

1978 年 12 月 18 日

第 三 版 序

本书于 1976 年出版, 1986 年又出版了修订本 (即第二版), 并多次印刷. 1990 年左右, 这本书已售罄. 许多朋友向作者建议, 最好将这本书增订出版或者能够重印. 这是对本书的爱护, 也是对作者的期望和鞭策.

从 1976 年到现在已经整整三十年了. 这三十年中编码理论迅猛发展. 先拿纠错码来说, 1998 年出版了 V. Pless 和 W. Huffman 主编的 *Handbook of Coding Theory*, 是两厚本共两千余页的巨著; M. Sudan 在 2002 年国际数学家大会上获 Nevanlinna 奖, 获奖工作包括他关于 Reed-Solomon 码列表译码的多项式时间算法这一突破性工作, 这是编码工作首次在国际数学家大会上获得大奖; 近年来又出现了十分活跃的量子码、空时码等等. 再拿密码来说, 1976 年提出了公钥密码以后, 出现了民间钻研密码的庞大队伍, 计算机科学也渗透到密码学中来, 密码学形成了庞大的学科, 从 1992 年和 1995 年分别出版的 G. Simmons 主编的 *Contemporary Cryptography* 和 A. Menezes 等人主编的 *Handbook of Applied Cryptography* 可见一斑. 因此要将本书再次修订, 使之包括近代编码的进展, 实在不是作者能力所及.

本书是一本基础性的读物. 前两章是学习编码理论所需要的代数知识, 它的特点是用尽量少的抽象数学概念和知识来阐述这些内容. 第三章介绍移位寄存器序列, 特别是线性移位寄存器序列的理论, 比较完整, 是进一步钻研非线性移位寄存器序列 (包括反馈序列及前馈序列) 的基础. 第四章介绍了几类重要的纠错码, 在此基础上可进入纠错码的近代发展. 因此再版这本书对打算进入编码或密码理论的读者应该是有益的.

作者非常感谢高等教育出版社的赵天夫、王丽萍等同志, 他们经过仔细的调研, 认真比较本书和其他同类著作, 并考虑读者的需要, 同意再版这本书. 这次再版, 除了校正了修订本的一些排印错误并对符号作了一些改进以外, 还对内容作了一些重要的修改. 特别, 在第三章 §8 中增加了序列的线性复杂度这一重要概念, 并利用这一概念简化了这一节里一个重要引理的证明 (这个简化证明是 1982 年瑞士高等工业学院的学生 U. Maurer 和 A. Viscardi 给出的), 而这个引理在解线性移位寄存器综合问题的 Berlekamp-Massey 迭代数算法的证明中起着关键作用. 其他的修改与增补, 在此不一一列举.

欢迎读者和专家对本书的批评和意见.

万哲先
2007 年 1 月

《非线性移位寄存器》序[①]

  近二十年来，由于反馈移位寄存器所产生的一些二元序列有许多重要应用，所以它的研究很受重视. 例如，在连续波雷达中可用作测距信号，在遥控系统中可用作遥控信号，在多址通信中可用作地址信号，在数字通信中可用作群同步信号，此外还可用作噪声源以及在保密通信中起加密作用等. 关于线性移位寄存器，由于有有效的代数工具对它进行分析，它的理论已相当完整. 至于非线性移位寄存器的理论，目前还很不成熟，这方面的结果，除戈龙布 (S. W. Golomb) 将他本人和他的一些合作者的几篇工作报告汇编成《移位寄存器序列》[*Shift Register Sequences* (Holden-Day, San Francisco, U. S. A., 1967)] 一书外，还散见于各杂志和报告中. 鉴于我国有关读者迫切要求这方面读物，特将各书刊上已发表的主要研究结果汇编成书，书中也包括我们自己工作中所得到的一些新成果和新证明. 本书出版后，希望能得到读者的批评指正.

作　者

1975 年

[①] 万哲先、戴宗铎、刘木兰、冯绪宁合著的《非线性移位寄存器》一书 1978 年由科学出版社出版.

Geometry of Classical Groups over Finite Fields: Preface[①]

Preface to the First Edition

This monograph is a comprehensive survey of the results obtained on the geometry of classical groups over finite fields mainly in the 1960s and early 1990s.

For the convenience of the readers I start with the affine geometry and projective geometry over finite fields in Chapters 1 and 2, respectively. Among other things, the affine classification of quadrics is included in Chapter 1, and conics and ovals are studied in detail in Chapter 2. From Chapter 3 and onwards the geometries of symplectic, pseudo-symplectic, unitary, and orthogonal groups are studied in succession. The book ends with two appendices, on the axiomatic projective geometry, and on polar spaces and finite generalized quadrangles, respectively.

Now I shall say a few words about the problems we are going to study in Chapters 3–7, and in addition give some historical remarks.

Let $\mathbb{F}_q$ be the finite field with q elements, where q is a power of a prime, $\mathbb{F}_q^{(n)}$ be the n-dimensional row vector space over $\mathbb{F}_q$ and $GL_n(\mathbb{F}_q)$ be the general linear group of degree n over $\mathbb{F}_q$. It is well-known that $GL_n(\mathbb{F}_q)$ has an action on $\mathbb{F}_q^{(n)}$ and an induced action on the subspaces of $\mathbb{F}_q^{(n)}$, which are thus subdivided into orbits under the action of $GL_n(\mathbb{F}_q^{(n)})$. It is natural to ask:

(i) How should the orbits be described?

[①] 万哲先所著的 *Geometry of Classical Groups over Finite Fields* 一书于 1993 年由瑞典 Studentlitteratur 出版社和英国 Chartwell-Bratt 出版社联合出版. 2002 年科学出版社出版了第二版.

(ii) How many orbits are there?

(iii) What are the lengths of the orbits?

(iv) What is the number of subspaces in any orbit contained in a given subspace?

The answers to these questions are classical and well-known for $GL_n(\mathbb{F}_q)$, but a natural question arises: If $GL_n(\mathbb{F}_q)$ is replaced by any one of the other classical groups: the symplectic group $Sp_{2\nu}(\mathbb{F}_q)$ (where $n = 2\nu$), the pseudo-symplectic group $Ps_{2\nu+\delta}(\mathbb{F}_q)$ (where q is even, $n = 2\nu + \delta$, and $\delta = 1$ or 2), the unitary group $U_n(\mathbb{F}_q)$ (where q is a square), and the orthogonal group $O_{2\nu+\delta}(\mathbb{F}_q)$ (where $n = 2\nu + \delta$ and $\delta = 0$, 1, or 2), then what will the answers of the four problems mentioned above be? This is what we shall analyse in Chapters 3–7.

In 1937 E. Witt studied problem (i) for the orthogonal group $O_n(F, S)$ defined by an $n \times n$ nonsingular symmetric matrix S over any field F of characteristic $\neq 2$. His famous theorem asserts that two subspaces P_1 and P_2 of $F^{(n)}$ belong to the same orbit under the action of $O_n(F, S)$ if and only if they have the same dimension and $P_1 S\, {}^t P_1$ and $P_2 S\, {}^t P_2$ are cogredient, where P_1 and P_2 also denote matrix representations of the subspaces P_1 and P_2, respectively. Later C. Arf, J. Dieudonné, L. K. Hua, et al. extended Witt's theorem to other classical groups.

In 1958 B. Segre studied problem (iii) for orthogonal groups over any finite field $\mathbb{F}_q$ but he restricted himself to consider only totally isotropic and totally singular subspaces where q is odd and even, respectively. He used the geometrical method and geometrical language. In his address at the International Congress of Mathematicians, Edinburgh, 1958, he announced his formula for the number of subspaces of a given dimension lying on a non-degenerate quadratic in the projective space over any finite field without any restriction on its characteristic.

In the mid-1960s problem (iii) was attacked by three of my students at that time, Z. Dai, X. Feng, and B. Yang, and myself. We obtained the closed formulas for the lengths of all the orbits of subspaces under the action of the symplectic, unitary, and orthogonal groups over finite fields. Our method is algebraic. Closed formulas for problem (iv), i.e., for the number of subspaces in any orbit contained in a given subspace were also obtained by myself in 1966.

In the early 1990s I studied problems (i) and (ii). Conditions satisfied by the numerical invariants characterizing the orbits of subspaces under the symplectic, unitary, and orthogonal groups over finite fields were obtained and then the number of orbits was computed. Problems (i)–(iii) for the pseudo-symplectic groups over finite fields of characteristic 2 were studied together with Y. Liu. Moreover, I also studied problem (iv) for the pseudo-symplectic group and problems (i)–(iv) for the singular symplectic, pseudo-symplectic, unitary, and orthogonal groups.

My interest in the geometry of classical groups over finite fields was arisen by block designs in the mid-1960s and by authentication codes in the early 1990s. By using it, many interesting block designs and authentication codes can be constructed. However, due to the limitation of the thickness of the book only some simple authentication codes constructed from the geometry of classical groups over finite fields are included in this monograph as examples. Using the geometry of classical groups over finite fields one can also construct projective codes with a few distinct weights and study the weight hierarchies of them; they are, however, not included in this monograph.

A large portion of the book is adopted from the lecture notes of a course entitled "Finite Geometry" which I gave at the Department of Information Theory, Lund University. I would like to express my sincere gratitude to Professor Rolf Johannesson who invited me to visit Lund and to give the course and encouraged me to write the present book. My visit in Lund is most fruitful and many of the results on the geometry of classical groups over finite fields that I obtained in Lund are compiled in the present book. The author is also deeply indebted to Mrs. Lena Månsson for her beautiful typesetting, careful and hard work, and most of all her patience and cooperation.

Zhexian Wan (Zhe-Xian Wan)
Lund, 1993

Preface to the Second Edition

As applications of the Anzahl theorems in the geometry of classical groups over finite fields some critical problems of finite vector spaces, Moor-Penrose generalized inverses of matrices over a finite field, and the representation of a form (bilinear, alternate, hermitian, or quadratic, etc.) by another form of the same kind over a finite field are added in this edition.

Zhexian Wan
Beijing, 2002

Geometry of Matrices: Preface[①]

The present monograph is a state of the art survey of the geometry of matrices. Professor L. K. Hua initiated the work in this area in the middle forties. In this geometry, the points of the space are a certain kind of matrices of a given size, and the four kinds of matrices studied by Hua are rectangular matrices, symmetric matrices, skew-symmetric matrices and hermitian matrices. To each such space there is associated a group of motions, and the aim of the study is then to characterize the group of motions in the space by as few geometric invariants as possible. At first, Professor Hua, relating to his study of the theory of functions of several complex variables, began studying the geometry of matrices of various types over the complex field. Later, he extended his results to the case when the basic field is not necessarily commutative, discovered that the invariant "adjacency" alone is sufficient to characterize the group of motions of the space, and applied his results to some problems in algebra and geometry. Professor Hua's pioneer work in the area has been followed by many mathematicians, and more general results have been obtained. I think it is now time to summarize all results obtained so far, and this has been my motivation for the present work.

In order to be as self-contained as possible the book covers some material of linear algebra over division rings in Chapter 1, which is necessary for later chapters. This chapter can also be read independently as an introduction to linear algebra over division rings. The fundamental theorems of the affine geometry and of the projective geometry over any division ring constitute the main contents of Chapter 2. In particular, Hua's beautiful theorem on

---

① 万哲先所著的 *Geometry of Matrices* 一书 1996 年由新加坡世界科学出版社出版.

semi-automorphisms of a division ring and its application to the fundamental theorem of the one-dimensional projective geometry over a division ring are included. Following these chapters, the geometry of rectangular matrices over any division ring, alternate matrices over any field, symmetric matrices over any field, and the geometry of hermitian matrices over any division ring which possesses an involution are discussed in detail in Chapters 3, 4, 5, and 6, respectively. Applications to problems in algebra, geometry, and graph theory are included throughout.

Finally, the author is indebted to Yangxian Wang and Mulan Liu for their helpful comments on the first draft of the book, to Rongquan Feng, Lei Hu, Xinwen Wu, and Zhanfei Zhou for their laborious typewriting, and to Lena Månsson for her beautiful improvement of the camera-ready copy.

Zhexian Wan

《有限典型群子空间的轨道生成的格》序①

我国典型群的研究,是华罗庚教授在 20 世纪 40 年代开创的,其特点是在几何背景的指导下,用矩阵方法研究典型群,此方法在典型群的结构和自同构的研究中很有成效,在 20 世纪中叶取得了丰硕的成果,受到国际同行们的重视,他们把以华罗庚为代表的典型群研究群体誉为典型群的"中国学派". 当时的研究成果多数汇集在《典型群》(华罗庚,万哲先著. 1963,上海科学技术出版社) 这部专著中.

后来,典型群的研究领域逐步扩大,万哲先与他的学生和合作者们对有限域上典型群几何学的理论和应用作了深入的研究,其应用所涉及的内容有:结合方案和区组设计、认证码、射影码和子空间轨道生成的格等. 关于有限域上典型群几何学理论方面的成果汇集在《有限域上典型群的几何学》(*Geometry of Classical Groups over Finite Fields*, 万哲先著, Studentlitteratur, Sweden/Chartwell-Bratt, United Kingdom, 1993) 这部专著中. 关于它对结合方案和区组设计的应用见《有限几何与不完全区组设计的一些研究》(万哲先,戴宗铎,冯绪宁,阳本傅著,1966,科学出版社) 这部专著. 另一些应用方面的成果散见近几年国内外有关的专业刊物.

本书讨论在有限域上的各种典型群作用下,由各个轨道或相同维数和秩的子空间生成的格. 当然,在一般线性群、辛群和酉群作用下,上述两种类型的格是一致的;而在正交群或伪辛群的作用下就需对这两种类型的格分别进行讨论. 在同类型的格中,首先研究不同格之间的包含关系;其次对给定的格中子空间的特性进行刻画;最后讨论所述格的几何性和计算它的特征多项式. 为了使本书的内容在阐述上系统完整,便于读者阅读,我们在第一章中介绍了格、几何格和特征多项式的一些基础知识,而在第二章到第十章中,按典

---

① 万哲先、霍元极《有限典型群子空间的轨道生成的格》一书 1997 年由科学出版社出版,2004 年出了第二版.

型群的通常顺序介绍了各种格的有关内容. 全书是用矩阵方法进行讨论和推导的. 我们认为这样处理比较具体直观, 便于读者学习参考.

  本书是我们和我们的合作者陈冬生、刘迎胜在这一领域中研究成果的系统介绍, 其中大部分已在国内或国际的学术刊物上发表, 也有一部分在国内或国际的一些学术会议上作过报告, 受到同行们的关注. 本书是在国家自然科学基金的资助下完成的, 刘嘉善编审对本书的出版给予大力支持, 在此一并致谢.

万哲先　霍元极

1995 年 8 月

第二版前言

  本书第一版出版后, 作者对书中涉及的有关问题作了进一步的探讨. 给出了辛空间、酉空间、正交空间和伪辛空间中子空间包含关系的矩阵表述及必要条件, 讨论了有限典型群作用下子空间轨道生成的格及同维数和秩的子空间生成格的几何性. 为使第二版所增加的内容与前面的 (第一版的内容) 衔接, 减少不必要的重复, 故对前面的内容作适当的调整和补充, 使本书更具有系统性. 本书第一版出版后, 发现了一些叙述上的不足和印刷上的错误, 在第二版中作了更正. 本书第二版增加的内容, 其中一部分已在国际刊物上发表, 也有一部分在一些学术会议上作过报告, 但大部分结果是在本书第二版中首次给出的, 例如第六章的 §6.5, §6.6 和 §6.8; 第七章的 §7.7, §7.8, §7.10 和 §7.11; 第八章的 §8.6, §8.7 和 §8.9; 第九章的 §9.6, §9.7 和 §9.9; 第十章的 §10.6, §10.7, §10.9 和 §10.10. 本书第二版也是在国家自然科学基金和河北省自然科学基金资助下完成的, 科学出版分社的责任编辑对本书出版给予大力帮助, 在此一并致谢.

万哲先　霍元极

2002 年 4 月 5 日

Lectures on Finite Fields and Galois Rings: Preface[①]

The present book is based on a course on finite fields I gave at Nankai University, Tienjin and a seminar on Galois rings I conducted at Suzhou University, Suzhou, both in 2002.

The first five chapters and Chapter 12 of the book are prerequisites for studying finite fields and Galois rings, respectively. They are prepared for students who had no background on abstract algebra, for instance, students in computer science or communication engineering. For those who already took a course on abstract algebra these chapters may be skipped. Chapter 6 to Chapter 11 are the main contents of finite fields; they are: structure theorems, automorphisms, norms and traces, various bases, factoring polynomials, constructing irreducible polynomials, and quadratic forms over (or of) finite fields. Galois rings are treated in Chapter 14, and Hensel's Lemma and Hensel lift, which are needed in studying Galois rings, are contained in Chapter 13.

It is helpful to treat finite fields and Galois rings in the same book. They have many similarities and their theories are parallel. Usually one can get some inspiration in Galois rings from finite fields. Moreover, it is also an economical way to study them simultaneously.

I hope this book can be used as a textbook for students in mathematics as well as also for those majoring in computer science or communication engineering. Of course, it can also be used for self-study.

---

① 万哲先 *Lectures on Finite Fields and Galois Rings* 一书 2003 年由新加坡世界科学出版社出版, 2006 年由世界图书出版公司北京公司重印出版.

People are interested in finite fields and Galois rings, mainly because they have important applications in science and technology; for instances: shift register sequences, algebraic coding theory, cryptography, design theory, algebraic system theory, etc. However, in order to make this book not too thick, these applications are not included.

In preparing the book I benefitted a lot from the existing literature on finite fields and Galois rings; in particular, the encyclopedia of Lidl and Niederreiter, the books of Jungnickel, McDonald, McEliece, and Menezes et al.

The author is most grateful to Dr. Qinghu Hou and the graduate students Mei Fu., Jun-Wei Guo, Qiu-Min Guo, Xia Jiang, De-Gang Liang, Chun-Lin Liu, Yan-Ping Mu, Yun Qin, Chao Wang, Ling-Ling Yang, Bao-Yin Zhang, Jing-Yu Zhao of the Center for Combinatorics of Nankai University who went through the tedious work of making a fair copy of the book. The author thanks Ms Jinling Chang who helped him with Latex problems and the compiling of the index. Finally, the author is also indebted to Professor Zongming Tang of Suzhou University who read the whole text and gave valuable comments.

Zhexian Wan
Beijing, 2003

Finite Fields and Galois Rings: Preface

The present book is a revision and addendum of a former one entitled Lectures on Finite Fields and Galois Rings, which will be referred as "Lectures" in the following. The "Lectures" is based on a course on finite fields I gave at Nankai University, Tianjin and a seminar on Galois rings I conducted at Suzhou University, Suzhou, both in 2002. It has been used as a textbook for students in mathematics as well as those majoring in computer science or communication engineering. Of course, it has also been used for self-study.

The first five chapters and Chapter 12 of the "Lectures" are prerequisites for studying finite fields and Galois rings, respectively. They are prepared for students who had no background on abstract algebra, for instance, students in computer science or communication engineering. For those who already took a course on abstract algebra these chapters may be skipped. Chapter 6 to Chapter 11 of the "Lectures" are the main contents of finite fields; they are: structure theorems, automorphisms, norms and traces, various bases, factoring polynomials, constructing irreducible polynomials, and quadratic forms over (or of) finite fields. Galois rings are treated in Chapter 14, and Hensel's Lemma and Hensel lift, which are needed in studying Galois rings, are contained in Chapter 13.

In the present book many typos of the "Lectures" are corrected, some changes are made for clarity, and many exercises are added. The most significant additions to the "Lectures" are a section on optimal normal basis (i.e., section 8.5) and an expansion of the original section 12.4 so that a simple proof of Theorem 8.34 is included.

People are interested in finite fields and Galois rings, mainly because they have important applications in science and technology; for instances: shift register sequences, algebraic coding theory, cryptography, design theory, algebraic system theory, etc. However, in order to make this book not too thick, these applications are not included.

In preparing the book I benefitted a lot from the existing literature on finite fields and Galois rings; in particular, the encyclopedia of Lidl and Niederreiter, the books of Jungnickel, McDonald, McEliece, and Menezes et als.

The author is most grateful to Professor Qinghu Hou of Nankai University and Drs. Yuping Deng, Ruoxia Du, Mei Fu, Jun-Wei Guo, Qiu-Min Guo, Xia Jiang, De-Gang Liang, Chun-Lin Liu, Yan-Ping Mu, Yun Qin, Chao Wang, Limin Yang, Ling-Ling Yang, Yiting Yang, Bao-Yin Zhang, Jing-Yu Zhao who went through the tedious work of making a fair copy of the book during their attending my lectures at Nankai University. The author thanks Ms Jinling Chang who helped him with Latex problems and the compiling of the index. Finally, the author is also indebted to Professor Zhongming Tang of Suzhou University who read the whole text and gave valuable comments.

Thanks are due to those readers who took the trouble to point out typos of the "Lectures".

Zhexian Wan
Beijing, 2010

怀念杨武之老师[①]

万哲先

杨武之老师是使我受益很多的一位老师,也是我非常敬重和经常怀念的一位老师.

1943—1944 年这一学年我在昆明联大附中读中学六年级. 学校的全称是西南联合大学师范学院附属中学, 现改名为云南师范大学附属中学. 联大附中实行六年一贯制, 中学六年级相当于高中三年级. 开学前, 我到学校注册的时候就听说, 我们联大附中的主任黄钰生教授 (西南联大师范学院院长, 兼附中主任), 为了加强我们班的教学, 聘请了好几位著名的专家来教我们班的课, 其中聘来教我们数学的是清华大学算学系主任兼西南联大算学系主任杨武之教授. 但我们班同学也担心, 大学教授会不会把中学数学看得太简单, 能不能把中学数学教好?

第一节数学课, 联大附中的教导主任魏泽馨老师陪着杨武之老师走进教室, 向我们介绍了杨老师. 魏老师告退后, 杨老师就开始讲课, 从远古时代人们由于生活和生产的需要, 结绳记数, 数数, 从而发现了自然数说起; 谈到后来为了把收获物分成一定的份数, 以便一人分一份, 从而发现了分数; 再谈到为了计算简便又创造了小数, 由于商业上借贷的需要又创造了负数; 再谈到由于建筑上需要正方形对角线的长和圆周长, 从而创造了无理数, 又引进了数轴, 等等. 杨老师用的语言通俗易懂, 讲解深入浅出, 举例生动有趣, 讲课引人入胜. 同学们像听故事一样听得津津有味, 觉得内容十分新颖, 没有一个同学不在专心听讲, 整个教室鸦雀无声, 同学们的担心一扫而光. 杨老师这一场关于数的起源和发展的开场白足足用了三节课, 接着自然地把讲课从数轴引到解析几何中的坐标系, 开始讲授解析几何. 杨老师所勾画的数的起源

[①] 原文刊登在《杨武之先生纪念文集》, 北京: 清华大学出版社, 1998, 129–132 页及《数学通报》, 1998 年第 10 期.

和发展的蓝图，我至今记忆犹新．

　　杨老师讲课的一个突出的特点是不断提问，几乎每节课，我们全班二十六个同学都有可能被提问一次．上面一段介绍的杨老师讲的数的起源和发展就伴随着无数次的提问．我体会，提问最大的好处是启发学生的思维．全班同学都跟着杨老师的提问在思考，因此课堂气氛非常活跃．学生通过自己的思考而获得知识，这样知识掌握得更为灵活，更为牢固．

　　杨老师一再教导我们，作图要力求准确，计算要力求正确，这成为我一直遵循的准则．他还在课堂上一再提醒我们不要犯中学生常犯的一些错误，如 $(a+b)^2 \neq a^2 + b^2$，而是 $(a+b)^2 = a^2 + 2ab + b^2$，等等．

　　杨老师还经常抓住机会帮助我们复习巩固以前学过的许多知识．有一次，他提问好几个同学，梯形面积公式是什么？都答不上来．他就耐心地从单位长度为边的正方形面积为 1 讲起，讲到矩形面积公式及其推导、平行四边形面积公式及其推导、三角形面积公式及其推导，以及梯形面积公式及其推导，最后讲到圆周长和圆面积的公式．当然伴随着他的讲解是一系列的提问．这样他用了不多的时间，就帮助我们复习巩固了平面几何中面积这一章．后来他再有机会提问平面图形的面积公式，同学们都能流利地回答出来．我也一直记得他这一段精彩的关于面积的复习课．

　　解析几何教材中有一段是利用移轴和转轴来化简二次曲线的方程．公式的推导对于中学六年级的学生来说，非常复杂．我在预习的时候，望而生畏，心想老师讲课时，一定会把这一段推导跳过去．但是出乎我的意料，杨老师讲这一段的时候，却一步步地、仔细地在黑板上推导．他在推导过程中，不断提问，问同学怎样算下去，启发我们的思维，纠正我们计算中的错误，并教我们怎样更简单地进行计算．在这以后，无论是在学习中还是在工作中，碰到复杂的计算，我都是不厌其烦地算到底．

　　上面提到的几点，难免挂一漏万，但是还有一点不能不提的，就是杨老师谆谆教诲我们的态度．他一心一意想把我们数学教好，想让我们把数学学好．他虽然是大学教授，社会地位已经很高了，但是他没有一点架子，总是平易近人．无论有什么问题去问他，他都耐心地解答．无论什么时候有事到他家里去找他，他都热情接待．

　　1948 年秋，杨老师从昆明来到北京清华大学任教．杨师母和除当时在美国的杨振宁先生以外的子女仍留在昆明. 他对于又回到久别的清华园是非常兴奋的. 这学期他教一门高等代数课和一门初等数论课. 当时我是初等数论课的助教, 和选课同学一起听他的课. 虽然大学里讲课已经没有那么多提问, 但我觉得, 和五年前在中学里听他的课一样, 仍是启发善诱, 引人入胜. 那时候他已经是五十多岁的人了, 一个人离开家在外面, 生活困难是很多的. 闲谈时, 我也劝他早点把师母和子女接到北京来团聚. 1948 年 12 月, 他飞回昆明, 把师母和子女接到上海, 等待迎接上海的解放. 后来, 他先后在同济大学和复旦大学任教.

    1953 年, 我从北京出差去上海, 到华山路他家里去看他. 当时他在复旦大学任教, 家住市内, 上课的几天乘车到市郊复旦大学住几天、上课. 这对于一位六十岁左右的老人来说, 是很不容易的. 他对清华园和清华大学以前的同事非常怀念, 在和我谈话中, 不断向我打听他们的情况. 他还鼓励我, 努力做研究, 要像开矿那样不断拓展, 要越开越深, 越开越大. 这些形象的话, 对我很有启发. 1976 年, 我又有机会从北京出差去上海, 那时他已经离开了我们. 我到他家去看杨师母, 还见到他们的小女儿杨振玉女士. 在和她们谈话中, 我说, 在我中学的最后一年很幸运有杨老师教我数学, 这对我的成长很重要; 而且我说, 我也常用杨老师的教诲去教我的学生. 杨师母向我说, 杨老师和她晚年最大的安慰就是他们的子女在学业上都很有成就 (用杨师母的话说, 就是都很用功, 都功课好).

1997 年 3 月 25 日

回忆母校联大附中[1]

中国科学院院士、1944 届校友　万哲先

云南师范大学附属中学的前身是西南联合大学师范学院附属中学. 她创办于 1940 年, 当时叫西南联合大学师范学院附设学校, 1940 年 11 月开学, 当时有小学六个年级, 初中三个年级. 1941 年增设高中一年级. 1942 年中学部和小学部分开, 中学部叫西南联合大学师范学院附属中学 (以下简称联大附中).

1942 年夏天我通过联大附中的入学考试, 荣幸地被录取为中学五年级插班生. 该年联大附中开始实行六年一贯制, 中学五年级相当于一般中学的高中二年级. 学校因为没有校舍, 11 月间才开学上课. 开始时是在昆明北门街南菁中学旧址的一个空场上露天上课, 一下雨就停课, 这样坚持了一个月左右, 借用南菁旧址的中法大学从南菁旧址迁出, 我们才有了教室. 我们在南菁旧址一直呆到 1943 年暑假前. 暑假后我们迁到钱局街校舍, 这个校舍是由一座破旧祠堂改建的, 我又在这个校舍里上了一年课. 我于 1944 年暑假前毕业, 总共在联大附中读了两年书. 这两年对我成长和发展影响很大. 为了纪念母校成立六十周年, 对这两年的往事作一些回忆.

联大附中主任由西南联大师范学院院长黄钰生教授兼任, 在附中校园里却经常见到他, 有时在找教员或学生谈话, 有时在课堂里讲课, 有时在学生课外活动的英诗班上讲英诗, 有时还在篮球或排球场边为学生的比赛鼓掌助兴, 等等. 他特别在选教员、建立优良的学风上, 花了很多的心血. 学校的老师大多是西南联大品学兼优的毕业生, 学识渊博而且富有钻研精神. 黄老师要找每位新教员谈话, 问他有什么困难并帮助解决, 交换对教育工作的看法, 并希

---

[1] 原文刊登于《情系三环 —— 庆祝西南联大附中建校 60 周年纪念文集》, 云南师大附中、云南师大附中校友会编, 内部资料, 2000, 242–252.

望他树立当一名优秀教师的理想. 1986 年他在天津担任天津市政协副主席、天津市图书馆馆长和天津市联合大学校长等要职. 那年我出差去天津, 抽空去拜访他. 他见了我, 第一件事就把他的夫人叶一帆老师的遗照拿给我看, 问我认不认识她. 我回答说, "这是叶先生, 教过我化学, 讲书非常清楚, 我得益很多." 接着他就说: "我对得起你们, 我给你们请的教员都是大学毕业的, 我把他们都看作自己的朋友." (注, 当时昆明所有中学的教员大学毕业的很少.) 我回答说: "您给我们请的老师不只是大学毕业的, 有的当时已是著名学者, 有的是当时大学毕业生中的佼佼者, 后来成为著名的学者. 附中的教育使我受用一生."

联大附中两年的教育是当时我能接受到的最好的中学教育. 教我们中学五年级语文的刘玚昌老师学识渊博, 他讲起课来, 引经据典, 丰富了我们的国学知识, 提高了我们对古文的理解力. 他还是我们班班主任, 经常和我们谈思想、谈志愿和谈做人. 他深受我班同学爱戴. 1946 年他在北京大学中文系任教时, 我班同学常常相约去看他. 后来他在武汉大学中文系任教授. 我们读中学六年级时, 采用的语文课本可说是一本中国历代哲学著作选编. 为了使我们能更好地理解课文, 黄钰生老师特别请来了西南联大哲学系的任继愈老师教我们语文. 他把艰深的哲学经典剖析给我们听, 使我们懂得了许多古代哲学思想. 任老师当时就是西南联大哲学系青年教师中的佼佼者, 后来成为我国著名的哲学家和宗教学家, 一代宗师. 他曾创办中国科学院哲学社会科学部宗教研究所并任所长, 后来又被委以国家图书馆馆长的重任.

中学五年级时教我们代数的是西南联大算学系毕业的邓汉英老师. 过去我学数学只满足于做习题, 但邓老师却还强调公式和定理的推导, 使我对数学的理解更进了一步, 逐渐对数学产生了兴趣. 黄钰生老师还专门请来西南联大算学系主任杨武之教授教我们中学六年级的 "解析几何" 和 "立体几何". 杨老师教学经验丰富, 教学效果极好. 我曾有撰文《怀念杨武之老师》, 就不再重复了. 黄钰生老师还请了西南联大算学系龙季和老师来担任联大附中的算学科主任. 龙老师为我们中学五年级开设 "向量分析" 选课, 并组织领导学生的数学课外活动小组 π 社. 这些都提高了我对数学的兴趣. 后来龙老师任广西大学教授, 创办了广西大学数学系并任系主任. 邓汉英老师则任南开大学数学系教授、系主任. 杨武之、龙季和、邓汉英三位老师都对我后来学数学很有影响,特别龙老师的影响更大.

黄钰生老师还请他的夫人叶一帆老师和西南联大化学系的黄新民老师先后来教我们化学课.他们都很注意实际应用,化学方程式在他们口中和笔下变成活生生的东西,使我们学起来很省力.后来,黄新民老师从事国防研究,很有贡献.我们中学六年级的英语教师由黄钰生老师亲自担任.他注重英语词组和英语习惯用法,给我们打下了很好的基础.我一直记得他充满感情地给我们讲英国作家莎士比亚的《裘力斯·恺撒》的选段和美国作家欧·亨利的《最后一片树叶》的情景.

中学六年级时教我们世界历史的是西南联大历史系的毕业生李道揆老师.李老师每节课都有一个题目,第一节课的题目是劳动创造了人类,后来我才知道这是对我们进行唯物史观的教育.李老师讲课引人入胜,史料在他嘴里变成了活生生的材料.我听他的课非常专心,仔细记笔记,下课后还重新整理,很工整地写下来.有一次他收了同学们的笔记本去检查,发现我重新整理的历史笔记,就很谦虚地对我说:不值得花这么多时间去整理,应该把这些时间花在读书上,并介绍我读几本世界历史.实际上,他讲课很有内容,也很新鲜,引起我对历史极大的兴趣,我多次动了中学毕业后考大学历史系的想法.李道揆老师后来成了美国史专家,在中国社会科学院美国史研究所任研究员,有很多论著.

我在联大附中读书的两年里,担任教导主任的是魏泽馨老师.他和黄钰生老师一样,是一心扑在教育事业上的教育家.他协助黄老师组织全校的教学,也和黄老师一样亲自到班上去上课,教一班语文.他有时还针对各班学生的特点和问题,到各班去做报告.我还清楚地记得,我在中学五年级时,他到我们班上来作怎样治学的报告,介绍一些大学问家的治学之道,并有的放矢地纠正我们读书中的问题,还特别指出傲气是学习中的一大障碍.新中国成立前夕,魏老师因支持学生运动,被捕入狱.新中国成立后,魏老师先后任长沙市教育局长、长沙联合大学校长等职.联大附中教过我的老师还很多,我都从他们那里受到教益.

联大附中除了重视课堂教学以外,还很重视组织学生的课外活动.学校里的体育运动(包括篮、排、足球等)非常活跃,音乐、戏剧、美术都很提倡.例如,常组织班际或校际的球赛,每学期都公演话剧,举行歌咏比赛、演讲比赛等. 这使学生生动活泼, 得到多方面的发展.

毕业前夕, 黄钰生老师等学校负责人对学生进行毕业教育也是难忘的. 特别因为我们是第一届毕业生, 对我们更是费尽心机. 我记得, 黄钰生老师在魏泽馨老师的协助下, 在我们中学六年级下学期, 请了昆明的专家学者来校演讲, 每月一次. 曾请了西南联大算学系教授江泽涵讲笛卡儿 (解析几何的发明人), 请了一位著名医学家来讲巴斯德 (细菌的发现者), 等等, 用这些科学巨人形象对我们熏陶. 黄钰生老师还约每位毕业生谈话, 根据每个学生的不足处提出应该努力的方向并对报考大学院系提出建议.

黄老师还请西南联大的一些教授, 出大学入学考试模拟题, 让我们试考, 发现了知识中的欠缺, 加以补救, 等等.

我想联大附中办学成功的经验是: 有有理想的而且精心要把学校办好的教育家, 组织了一个高水平的教员班子, 招收了一批合格的学生.

我从联大附中毕业已经 56 年了, 联大附中对我的教育和联大附中的学生生活我一直记忆犹新. 我对联大附中怀有深厚的感情. 1982 年我去昆明开会, 到昆明的第二天就到附中校舍, 到了校门口, 喜出望外, 东张西望, 不知向谁打听, 打听谁. 碰巧姜为藩校长来到校门口, 问我找什么人. 我操着昆明话说, 我就是找这所学校, 我是这所学校的毕业生. 姜校长就约我到他办公室去谈话. 从谈话中了解到, 云南师大附中不仅是省重点中学, 而且是云南省首屈一指的学校, 为国家培养了大量的优秀毕业生, 我由衷地感到高兴. 我希望今后有机会再去昆明, 再去探望云南师大附中, 届时云南师大附中一定会更上一层楼.

Hsiao-Fu Tuan, 1914–2005

Zhexian Wan

Professor Hsiao-Fu Tuan, a leading algebraist in China and the honorable member of the Editorial Board of Algebra Colloquium, passed away on 6 February 2005 at the age of ninety-one.

Professor Tuan was born on 29 July 1914. He received his Bachelor's degree at Tsinghua University in 1936, his Master's degree at Toronto University in 1941, and his Ph. D. at Princeton University in 1943. He was a post-doctorate at Princeton University from 1943 to 1945, and the assistant to Hermann Weyl at the Institute for Advanced Study, Princeton from 1945 to 1946. After his return to China in 1946, he became a professor of Tsinghua University and subsequently at Peking University in 1952. In 1955, he was elected to be a Member (Academician) of the Chinese Academy of Sciences.

Professor Tuan was known worldwide for his contributions in finite groups and algebraic Lie groups. He is a foremost group theorist and the founder of the representation theory of groups in China. The modular representation theory of finite groups was established by Richard Brauer in 1935, and Tuan started to work on this subject in 1940 and proved some basic facts by himself or in collaboration with Brauer. Among his contributions, he determined the structure of a class of simple groups and linear groups, proved the completeness of the table of L.E. Dickson of finite simple groups of order at most 10,000, and established three important lemmas, which are later called the "Brauer–Tuan–Stanton principle", "Brauer–Tuan's separation principle of blocks of characters" and "Brauer–Tuan theorem". More than fifty years have elapsed, but these results are still shinning in the realm of group theory, because on the one hand, they concern the basic theory of finite groups and are the starting point of later developments, and on the other hand, later results cannot avoid or replace them. Tuan's results are elaborated in detail in many monographs and textbooks of the representation theory of finite groups, and are quoted extensively by group theorists in the world. The famous monograph "Representation Theory of Finite Groups" by Walter Feit is a case in point.

Professor Tuan made also remarkable contributions in algebraic Lie groups. An algebraic Lie group over the complex field is a complex matrix group, the coefficients of its matrices satisfying a system of algebraic equations. In 1943, Claudy Chevalley provided the definition of the replica of a matrix by the tensor invariants of the matrix and then gave the definition of the algebraic Lie algebra as a subalgebra of the $n \times n$ matrix algebra over a field of characteristic 0 using replicas of matrices. Then Chevalley and Tuan proved the fundamental theorem of algebraic Lie groups: "The Lie algebra of every algebraic Lie group is an algebraic Lie algebra and every algebraic Lie algebra over the complex field is the Lie algebra of an algebraic Lie group." Armand Borel pointed out that the work of Chevalley and Tuan of extending algebraic Lie groups to any field of characteristic 0 by Lie algebras is the overture of the birth of the general theory of linear algebraic groups.

Professor Tuan studied also stream ciphers and suggested some algorithm which facilitates the computation considerably.

Due to his remarkable scientific achievements, Professor Tuan was awarded the Scientific and Technological Progress Prize of the Ho–Leung–Ho–Lee Foundation in 2004.

Professor Tuan held the position of the chairman of the Department of Mathematics of Tsinghua University and the Department of Mathematics and Mechanics of Peking University for thirty four years. He was the main organizer and promoter in the development of algebra in China, and established a good style of study and created a cordial and friendly atmosphere. He made every effort to encourage Chinese algebraists to establish international contacts. Many leading mathematicians abroad paid visit to China at his invitation or under his sponsorship. He also organized quite a few important international conferences, notably the International Symposium on Group Theory, Beijing, 27 August — 8 September, 1984, the proceedings of which was published by Springer-Verlag. In his early days of teaching at Tsinghua University, Xihua Cao, Shisun Ding, Kencheng Zeng and myself were his students and were much influenced by him. After he moved to Peking University, among his students working in algebra are Guangyu Shen, Yichao Xu, Yizhong Lan, Daoji Meng, Caihui Lu, and in particular, working in group theory, E-fang Wang, Zhongmu Chen, Shengming Shi, Jiawei Hong, Huiling Li, Mingyao Xu, Jiping Zhang, Jie Wang, and others. The students of his students are numerous. As Charles W. Curtis said, "Upon his return to China, Tuan founded a school for research in modular representation theory and the structure of finite groups at Peking University. Under his leadership, it became a center for research, and is flourishing today".

深深怀念陈省身先生[①]

万哲先

  1946–1947 学年度的第二学期, 清华大学教授陈省身先生回到清华园, 给大学生开设《高等几何》课. 我当时是清华大学算学系三年级学生, 选修了这门课. 开学前, 系里通知选这门课的同学去陈先生家中和他见面. 见面后, 他一一地问我们的姓名, 询问过去学习的情况, 态度和蔼可亲. 为了解我们对基础知识掌握的程度, 他还提一些问题, 问题由浅入深, 步步深入, 要求我们回答确切, 不许含糊. 我们答不上来的时候, 他就循循善诱地启发我们回答. 通过短短两小时的谈话, 他对我们同学的情况有了初步了解, 为他的讲课作了准备, 讲课时做到有的放矢. 这次谈话也使我发现自己学习的一些不足之处, 回到宿舍, 赶快翻书, 思考演算, 复习巩固.

  谈话结束前, 他简短地介绍了他要把《高等几何》课分成《高等几何 (上)》和《高等几何 (下)》两门课程来讲授.《高等几何 (上)》讲授 Erlangen 纲领,《高等几何 (下)》讲授拓扑学. 这次谈话是我第一次见到陈省身老师, 当时他关于 Gauss-Bonnet 公式的内蕴证明和他引入的复向量丛的示性类 (后来被称作陈省身示性类) 已开始闻名于世, 他已是举世闻名的大数学家. 我们能碰到一位大数学家来教我们基础课, 确实幸运、荣幸.

  陈老师讲课内容新而且观点高. 听他的课, 学生受益良多. Erlangen 纲领我是第一次从陈老师那里听到的.《高等几何》课过去主要讲射影几何, 内容相当古老. 在《高等几何 (上)》这门课里, 他用 Erlangen 纲领把射影几何、仿射几何、非欧几何以及欧氏几何等分支串起来讲授, 同学们听了觉得十分新鲜、有趣, 看到几何学中居然有如此美妙的纲领, 开阔了视野, 加深了对数学的理解, 提高了对数学的兴趣. 在最后一节课里讲了计数几何, 内容十分新颖, 很吸引同学兴趣. 在《高等几何 (下)》这门课里他讲的是拓扑学. 拓扑学

[①] 原文刊登于《陈省身与中国数学》, 八方文化创作室, 新加坡, 2007, 109–112.

在当时刚刚兴起不久,是当时最时髦的数学科目. 许多数学家都预见到,这门新科目对数学的发展会有深远的影响. 陈老师把它放到大学生的课程中讲授在当时是一个创举,目的是想尽早把学生带到数学研究的第一线. 因为授课时间短,我们同学起点低,他只讲了一般拓扑学和同调论.

陈老师讲课十分清晰,讲课技巧很高,语言简明扼要,讲话句句重要,几乎没有一句多余的话. 每节课前我们都怀着极高的兴趣走进教室,听课时聚精会神,十分专注,跟着他进入美妙的数学仙境,下课后还不断回味. 听他的课真是极高的享受. 他备课极为认真负责,事先写好授课笔记,上课时板书非常清楚,还经常画些插图,举些例子来阐明抽象的数学理论和启发学生的思维. 他讲授的《Erlangen 纲领》和《拓扑学》两门课,我都记有课堂笔记,我一直把它们视为至宝. 可惜《Erlangen 纲领》一课的笔记在"文化大革命"中遗失,《拓扑学》的笔记一直保留到现在. 陈老师 2004 年初不幸去世后,我把它赠给了清华大学图书馆,希望他们永久保存. 记得陈老师给我们讲拓扑学课结束时曾说过,目前还没有一本拓扑学的入门书. 他准备把他的授课笔记整理成书出版,我一直在盼着这本书的出版,但他科研太忙,一直没能抽出时间来把这份授课笔记整理出版.

陈老师重视课下的作业,每周都布置习题给学生做. 有的习题是要运用课堂里学的理论来解,有的习题则是为了弥补以前学习之不足. 我们做了都很有收获. 学生交上来的作业由他自己认真批改,我们得益很多.

陈老师回到清华园不久,立即组织了算学系的(综合)讨论班,每周一次,由他自己带头讲第一讲,接着段学复老师、赵访熊老师、吴文俊先生、徐利治先生、吴光磊先生等等,一个接一个地演讲,系里的学术气氛立即活跃了起来.

可惜陈老师这次在清华园只待了一个学期. 1947 年 6 月,就离开北京去南京了. 因为当时他正在负责南京中央研究院数学研究所的筹建工作. 从那以后,直到 1972 年他从美国回国访问之前,我一直没有再见到过他.

我在大学期间能听到陈老师这样一位大数学家讲基础课是很幸运的,我是非常感激的. 后来我在大数学家华罗庚老师指导下从事代数学的研究,随后我又开展组合论的研究,陈老师的课给我打下的几何基础都一直在起作用.

1972 年以后,他不断回国访问. 他对祖国和以前的师友怀有深厚的感情,对祖国数学的发展非常关切, 热情希望为我国数学的发展贡献他的力量, 使我国早日成为数学大国和数学强国. 1980 年他倡议我国举办微分几何和微分方程国际学术讨论会 (简称双微会议), 两年一次, 每次都邀请国际上一流的数学大家作系统的学术演讲. 他还建议段学复教授仿照双微会议的模式举办群论国际学术讨论会, 等等. 这些活动对我国数学的发展, 年轻人才的成长起了重要的促进作用. 1985 年他更创办了南开数学研究所. 现在南开数学研究所已成为我国数学乃至国际数学的一个重要研究中心. 没想到陈老师竟在这时离我们而去, 陈老师的数学成就是永恒的, 陈省身数学研究所永远会屹立在东方. 陈老师对中国数学发展的贡献是巨大的, 我们将永远深深怀念他.

中国的代数学

万哲先

"中国的代数学"这个题目太大了,很难在一篇短文里把中国的代数学说得面面俱到. 所以我在此文中略去了代数学的一些边缘学科, 如代数数论、代数几何、Lie 群等,环论和半群也从略, "文化大革命"后的工作也谈得非常简略.

2006 年万哲先院士在全国代数会议上

古代

先举出三项我国古代在代数学方面的重要成就:

(1) 孙子的孙子定理(《孙子算经》, 该书成书于公元 200—499 年). 这个定理在西方称为"中国剩余定理".

---

① 原文刊登于《中国数学发展》,数学与人文,第一辑,北京: 高等教育出版社, 2010, 275-283.

(2) 秦九韶在 1247 年给出了多项式方程的数值解法. 五百多年后, 西方数学家 W. Horner 在 1819 年又独立地发现了这一方法.

(3) 杨辉 1261 年独立于西方发现了 Pascal 三角形.

曾炯 (1898–1940)

曾炯教授

曾炯 (字炯之) 是在抽象代数发展的早期进入抽象代数学领域, 并作出重要贡献的数学家, 更是我国从事抽象代数学研究的第一人. 他 1929 年入当时国际数学中心德国哥廷根大学, 在迄今为止最伟大的女数学家 E. Noether 指导下攻读抽象代数学, 1934 年获博士学位. 接着到德国汉堡大学著名数学家 E. Artin 那里进修. Artin 和 Noether 被认为是抽象代数学的两位鼻祖, 曾炯是唯一的一个曾受教于他们两位的中国代数学家. 1935 年 7 月曾炯回国, 在浙江大学任副教授, 1937 年被天津北洋大学聘为教授. 抗日战争爆发后, 曾炯随北洋大学西迁, 先后在西安、城固等地任教, 1939 年到西昌技艺专科学校任教. 1940 年 11 月因胃穿孔大量出血, 而西昌又缺医少药, 曾炯不治逝世.

曾炯在代数上的贡献可以概括成三条定理, 其中两条定理是关于可除代数的.

设 F 为域, D 叫做 F 上的可除代数, 如果

(1) D 是除环;

(2) D 是 F 上的有限维向量空间;

(3) 对任意 $a \in F$ 和 $u, v \in D$, $a(uv) = (au)v = u(av)$.

如果更假设 D 的中心是 F, D 就叫做 F 上的中心可除代数.

定理 1 (1933) 设 Ω 是代数闭域, X 是未定元, 那么 $\Omega(X)$ 上的中心可除代数只有 $\Omega(X)$ 自己. 更一般地, 设 K 是 $\Omega(X)$ 的有限次扩域, 那么 K 上的中心可除代数也只有 K 自己.

定理 2 (1934) 设 P 是实闭域, 而 K 是 $P(X)$ 的有限次代数扩域, 那么 K 上的中心可除代数除去 K 之外, 最多还有一个, 其指数为 2.

上面两个定理分别是"复数域上的中心可除代数只有它自己"和"实数域上的中心可除代数除去它自己之外只有实四元数除环"这两个经典结果的漂亮推广.

曾炯还引进了 C_i 域的概念. 域 F 称为 C_i 域, 如果对任意正整数 d 及 F 上任一 n 元 d 次齐次多项式 $f(x_1,\cdots,x_n)$, 若 $n > d^i$ $(i > 0)$, 则 $f(x_1,\cdots,x_n) = 0$ 必在 F^{d^i} 中有非全零解.

定理 3 (1936)   设 Ω 是代数闭域, 那么 $\Omega(x_1,\cdots,x_n)$ 为 C_n 域.

S. Lang 于 1951 年又重新得到这个定理, 在文献中这个定理被称为 "曾-Lang 定理". 这个定理是关于超越扩张的 Brauer 群研究的基础, 同时对 Artin-Schreier 形式实域上的二次型理论有重要应用, 因此被许多文献引用.

华罗庚 (1910–1985)

前面已经介绍, 曾炯在代数学上有重要贡献, 可惜他英年早逝, 工作中断. 对我国代数学的发展有重要影响的首推华罗庚. 华罗庚在数学上的贡献是多方面的, 在代数学上的贡献也是多方面的, 有 p 群、矩阵几何、典型群、除环等.

清华大学算学系合影 (1934). 前排左二起: 唐培经、赵访熊、郑之蕃、杨武之、周鸿经、华罗庚; 中排有: 陈省身 (左一)、段学复 (左四)

1938 年华罗庚在昆明西南联合大学组织了有限群讨论班, 参加的人除华罗庚本人外, 有徐贤修、段学复、樊䶮、孙本旺等. 这个讨论班因 1939 年的英庚款考试而中断. 在短短的一年多时间里, 这个讨论班的参加者写出了好几篇很有意义的关于有限群的论文,特别是 p 群的论文.

1939 年华罗庚、陈省身和王竹溪在昆明西南联大开了连续群讨论班,由他们三人轮流主讲. 这个讨论班当时是很先进的, 苏联数学家 L. S. Pontrjajin 的名著《连续群》刚刚出版. 一位听讲的学长告诉我说, 三位先生的演讲一次比一次精彩, 听的人受益很多. 连续群在华罗庚、陈省身后来的工作中起的作用是十分显著的.

顺便在这里提一下, 华罗庚、陈省身和王竹溪等更早一次的竞赛是在 1932—1933 学年度. 那一年熊庆来在清华大学开设高等分析课. 全班一共五个学生: 华罗庚、陈省身、许宝騄、柯召和王竹溪. 许、柯是本科生, 陈是研究生, 华是助理, 王是物理系高才生. 熊庆来采用 E. Goursat 著的 *Cours d'Analyse* 作为教材, 这本书上的习题很难, 熊庆来选了书上的一些题留作作业. 班上的五位学生便展开了做题竞赛, 今天你做出一题来, 明天我做出一题来, 后天他又做出一题来, 结果他们五位把 Goursat 书上的习题一题不漏地全做了出来. 大家在做题竞赛中前进, 水平突飞猛进. 我们应该感谢他们五位前辈没有出 Goursat 书上习题的解答书.

华罗庚在除环上有三条非常漂亮的定理. 设 D 是除环, $\sigma: D \to D$ 是双射, 并设 σ 满足条件 $\sigma(a+b) = \sigma(a) + \sigma(b)$ 对任意的 $a, b \in D$. 若 σ 还满足条件 $\sigma(ab) = \sigma(a)\sigma(b)$ 对任意的 $a, b \in D$, 则 σ 叫 D 的自同构; 若 σ 还满足条件 $\sigma(ab) = \sigma(b)\sigma(a)$ 对任意的 $a, b \in D$, 则 σ 叫 D 的反自同构; 若 σ 还满足条件 $\sigma(a^2) = \sigma(a)^2$ 和 $\sigma(aba) = \sigma(a)\sigma(b)\sigma(a)$ 对任意的 $a, b \in D$, 则 σ 叫 D 的半自同构.

定理 1 (1949) 除环的半自同构一定是自同构或反自同构.

也许有人会有疑问, 是不是华罗庚定义了半自同构然后作了一个逻辑游戏推出定理 1? 不是的. 华罗庚最反对把做数学当成逻辑游戏. 在 1949 年以前, 有数学家企图证明可除代数上的一维射影几何的基本定理, 引入了可除代数的半自同构这个概念, 华罗庚洞察到除环的半自同构必为自同构或反自同构是证明除环上一维射影几何基本定理的关键. 他先证明了定理 1, 然后推出了除环上一维射影几何的基本定理.

除环上一维射影几何的基本定理 设 D 为除环, 特征不为 2. 令 $PG^l(1, D) = D \cup \{\infty\}$. 那么 $PG^l(1, D)$ 到自身的双射, 如将调和点列变到调和点列, 必

为形状
$$x \to (a+x^\sigma c)^{-1}(b+x^\sigma d),$$
其中 σ 为 D 之自同构或反自同构, 而 $\begin{pmatrix} a & b \\ c & d \end{pmatrix} \in GL_2(D)$.

所谓 x_1, x_2, x_3, x_4 成调和点列, 是说
$$(x_2-x_4)^{-1}(x_2-x_3)(x_1-x_3)^{-1}(x_1-x_4) = -1.$$

定理 2 设 D 为除环, Z 为 D 的中心. 再设 D_0 是 D 的真子除环. 如果 $a^{-1}D_0 a \subseteq D_0$ 对任意 $a \in D^*$, 那么 $D_0 \subseteq Z$.

这个定理最初是 H. Cartan 利用他关于可除代数的 Galois 理论对可除代数证明的, 后来华罗庚和 R. Brauer 分别独立地对除环证明, 所以定理 2 被叫做 "Cartan-Brauer-Hua 定理".

定理 3 (1950) 设 D 是除环但不是域, 那么 D^* 是不可解群.

华罗庚在 20 世纪 40 年代初因研究多元复变函数论, 而开始研究典型域, 从而开创了矩阵几何的研究. 研究的主要问题是用尽可能少的不变量来刻画矩阵空间的运动群. 1945–1947 年他发表了一系列复数域上矩阵几何的论文, 1948 年以后他又将复数域上的矩阵几何推广到任意域, 甚至任意除环上, 并指出仅用保持粘切这一不变量即可刻画运动群, 这一结果被称为矩阵几何的基本定理. 1948 年他证明了特征不为 2 的对称矩阵几何的基本定理, 1951 年他又证明了除环上长方阵几何的基本定理.

20 世纪 60 年代初, 华罗庚的学生万哲先 (即本文作者) 又组织力量从事域和除环上矩阵几何的研究. 如, 1966 年刘木兰证明了任意特征的域上交错矩阵几何的基本定理. 矩阵几何的研究因 "文化大革命" 而中断. 到 20 世纪 90 年代初, 万哲先证明了任意特征的域上的对称矩阵几何的基本定理和带对合的除环上埃尔米特矩阵几何的基本定理, 这样将矩阵几何推广到任意域或体上的工作已趋完整. 在此基础上, 1996 年万哲先出版了专著《矩阵几何》, 以此来怀念华罗庚教授.

华罗庚关于典型群的工作是系统的、大量的, 涉及典型群的结构、自同构和表示论, 有许多追随者 (如万哲先、王仰贤、严士健、任宏硕、李尊贤、李福安、吴小龙、游宏、陈宇等), 影响很大, 被国外同行誉为 "典型群的中

国学派". 详细结果见华罗庚、万哲先的《典型群》(1963) 及华罗庚的《多复变数函数论中的典型域的调和分析》(1958) 两本专著, 这里就不一一列举了.

段学复 (1914–2005)

段学复院士

另一位对我国代数学的发展有重要影响的数学家是段学复, 他的代数工作也是多方面的, 有 p 群、有限群的模表示论、代数 Lie 群和代数 Lie 代数、移位寄存器序列 (= 递归序列) 等. 下面着重介绍段学复在有限群模表示论上的工作.

设 G 是有限群, $\psi: G \to GL_n(\mathbb{C})$ 是群同态, ψ 叫做 G 的群表示. 群的表示论是 20 世纪初由 F. Frobenius 和 W. Burnside 开创的, 其目的主要是用来研究有限群的结构. 阶为 $p^\alpha q^\beta$ (p, q 为相异素数) 的有限群皆可解, 这一漂亮的定理就是当时 Burnside 用群表示论证明的. 后来发现群表示论在理论物理中有重要应用, 更引起许多学者的重视和研究. 有限群上的函数 $g \to \mathrm{Tr}(\psi(g))$ 叫做 G 的表示 ψ 的特征标, 它是 I. Schur 引进的. 群指标的引进简化并推进了有限群表示论的研究.

设 p 是素数, $p \,|\, |G|$, 那么同态 $G \to GL_n(\overline{\mathbb{F}}_p)$ 就叫 G 的模表示, 这里 $\overline{\mathbb{F}}_p$ 是 p 元域 $\mathbb{F}_p$ 的代数闭包. R. Brauer 从 1935 年开始对有限群的模表示进行系统而深入的研究, 被认为是有限群模表示论的创始人. 段学复从 1940 年起参加到这一研究中去, 是除 Brauer 之外, 在有限群模表示论早期的发展中作出重要贡献的数学家之一.

利用有限群的模表示可将特征标分成块, 产生了特征标的块理论, 它成为探究群的结构的主要工具, 段学复和 Brauer 一起建立了特征标块的三个基本引理, 分别被人们称为 "Brauer-Tuan-Stanton 原则" "Brauer-Tuan 指标块分离原则" 和 "Brauer-Tuan 定理". 几十年过去了, 这些工作并未失去光彩, 而被详细写入 W. Feit 的名著 *Representation Theory of Finite Groups*, 并为群论工作者广泛引用.

段学复还运用有限群模表示论来研究有限线性群和有限单群. 例如,

(1) 对 pq 阶的线性群, 这里 p 是素数而 $(p, q) = 1$, 当维数 $\leqslant (2p+1)/3$

时, 确定了它们的构造;

(2) 证明了 L. E. Dickson 所著 *Linear Groups* 一书中的单群表直到 10000 阶是完全的.

递归序列和代数编码

设 F 是域, F 上的序列

$$a_0, a_1, a_2, \cdots \quad (a_i \in F,\ i = 0, 1, 2, \cdots)$$

叫递归序列, 如果有 $f(x_1, x_2, \cdots, x_n) \in F(x_1, x_2, \cdots, x_n)$ 使

$$a_k = f(a_{k-n}, \cdots, a_{k-2}, a_{k-1}), \quad \forall k = n,\ n+1,\ n+2, \cdots.$$

如果 f 是线性式, 则上式称为线性递归序列.

1971 年中国科学院数学研究所和北京大学数学力学系分头接触到线性递归序列的一个问题, 随后段学复和万哲先独立地提出了解这个问题的算法. 我国代数学家开始了解到代数可应用于递归序列和与之相关联的编码理论, 十分兴奋. 通过与通信工程师的接触, 他们还了解到那时通信工程不只是用到连续性数学, 如傅里叶变换等, 而且愈来愈需要离散数学, 特别是有限域的理论, 在通信中起着越来越重要的作用.

1972 年在段学复倡导下北京大学率先在国内举办多期编码方面的短训班, 每期两年, 为从事通信工程技术的人普及代数知识和编码知识, 中国科学院数学研究所的人也参加讲课. 随后, 四川大学、中国科学技术大学、中国科学院研究生院等院校也相继举办类似的短训班, 这些短训班为在我国普及代数和编码知识作出了重要贡献.

另一方面, 从 1972 年春开始, 万哲先在中国科学院数学研究所组织编码讨论班, 介绍国际上编码学和密码学的新发展, 并和他以前的学生戴宗铎、刘木兰、冯绪宁开展递归序列的研究. 他们对 M 序列的构造, 以及对 M 序列的前馈序列的分析与综合均有系统的工作. 当时在合肥中国科技大学工作的, 他以前的学生林秀鼎、冯克勤也到北京来参加编码讨论班和研究工作. 参加这个讨论班的还有北京大学数学力学系、中国科学院计算技术研究所、邮电科学研究院和北京邮电学院等院校的同志们. 1976 年万哲先的《代数和编码》一书出版 (2007 年出版了第三版), 1978 年他和上面提到的戴、刘、冯的

专著《非线性移位寄存器》出版. 这两本书在国内已成为相关领域的经典文献. 中国科学院数学研究所的编码讨论班一直延续到1977年, 戴宗铎、刘木兰、冯克勤等人的关于编码学和密码学的研究工作一直延续到现在.

在代数编码方面, 冯克勤以前的学生邢朝平作出了重要贡献. 1998年他与Niederreiter合作, 用代数几何构作了一些码, 打破了Gilbert界. 近年来冯克勤在量子码、空时码上均有有意义的工作.

"文化大革命"以后

1976年"文化大革命"结束后, 我国数学家采用请进来和派出去的办法, 一方面努力发扬我国数学的传统并使之与国际接轨, 另一方面又努力学习国外数学的新成就, 开展一些新分支的研究. 这样, 我国代数学除了群论(包括典型群和有限群模表示论)、环论(包括结合环和非结合环, 如 Lie 代数)等已有分支以外, 又开展了许多新分支的研究, 有些新分支是从原有的分支延伸出来的, 如从典型群到代数 K 论、典型群的几何学等, 从环论走向模论、同调代数、Hopf 代数和代数表示论等, 从有限维 Lie 代数走向无限维 Lie 代数和超 Lie 代数等; 也有的分支是从头开始的, 如线性代数群的表示论、交换代数等. 我国许多青年代数学家分别在这些分支中做出了优秀的工作, 这是值得大书特书的, 但是限于篇幅和我的知识面, 我只举典型群、有限群模表示论和线性代数群表示论为例, 简述如下.

有限域上典型群的几何学是典型群研究的继续和发展. 设有限域 $\mathbb{F}_q$ 上 n 维向量空间 $\mathbb{F}_q^n$ 上作用着一个 n 级典型群, 它可以是一般线性群 $GL_n(\mathbb{F}_q)$, 辛群 $Sp_{2\nu}(\mathbb{F}_q)$ $(n=2\nu)$, 酉群 $U_n(\mathbb{F}_{q^2})$ 或正交群 $O_n(\mathbb{F}_q)$. 这时, $\mathbb{F}_q^n$ 的子空间就分成一些轨道, 我们可以提出以下这些问题: (i) 两个子空间何时属于同一轨道? (ii) 总共有多少条轨道? (iii) 每条轨道的长度是多少? (iv) 一条轨道中共有多少个子空间包含一个给定的子空间? 1963年到1964年, 万哲先带领他的学生戴宗铎、冯绪宁、阳本傅给出了问题 (iii) 的完整答案, 还将结果应用到区组设计上, 并于1966年出版了专著《有限几何与不完全区组设计的一些研究》. 随后因为"文化大革命", 工作中断. 1990年左右万哲先又回来研究其余的问题, 给出了完整答案. 在此基础上, 万哲先撰写了专著 *Geometry of Classical Groups over Finite Fields*, 1993年出版, 2002年出了第二版.

万哲先还将有限域上典型群的几何学应用到众多领域和问题中, 如结合方案和区组设计、认证码、子空间轨道生成的格、型表型问题、纠错码、临界问题、强正则图等. 戴宗铎和冯克勤又分别将它用于广义逆和量子码. 曾肯成带领他的学生李尚志、查建国研究典型群的极大子群, 取得了丰硕成果. 在此基础上, 李尚志撰写了专著《典型群的子群结构》.

张继平教授

段学复关于有限群模表示论的工作由他的学生继续和发展. 1989 年张继平利用有限群模表示论和有限单群的分类, 改进了段学复关于复线性群的结果, 即将复线性群的维数 $\leqslant (2p+1)/3$ 改进为 $\leqslant p-1$. 这样, 他就解决了著名的 Brauer 第 39 问题和第 40 问题. 张继平还有其他的重要工作, 例如:

(1) p 块理论是有限群模表示论的核心, 亏数 0 的 p 块的存在性是模表示论的重要公开问题. 在有限群的 Sylow 子群循环的情况下, 他给出了亏数 0 的 p 块存在的第一个充要条件.

(2) 他创立和发展了群的算术理论. 利用群的算术理论, 他解决了 Huppert 级数猜想和 Markel 猜想等著名难题.

此外, 樊恽独立地在武汉也开展了模表示论的研究, 得到一些很好的成果. 例如, 他解决了任意系数域上幂零块的结构理论及几个相关问题, 与 M. Broué 和 L. Puig 合作完成了带幂零系数扩张的块的结构理论.

在线性代数群的表示论方面, 席南华的工作很出色. 例如,

(1) 他对于仿射 A 型 Weyl 群, 证明了 Lusztig 关于基环的猜想.

(2) 他确定了 Deligne-Langlands 关于仿射 Hecke 代数的猜想成立的充要条件.

席南华师从华东师范大学时俭益, 时俭益在研究仿射 Weyl 群胞腔方面有许多很好的成果, 特别在 Warwick 的博士论文 *The Kazhdan-Lusztig cells in certain affine Weyl groups* 中解决了 $\widetilde{A}_n$ 型仿射 Weyl 群的胞腔分解. 华东师范大学的王建磐在代数群的模表示和量子群上有很好的工作, 例如他与 Parshall 合作, 用量子函数代数的余模的观点对量子线性群的表示进行了深入系统的研究, 取得一系列成果. 华东师范大学代数群这一研究方向是曹锡

华倡导的.

从左至右: 肖杰、王建磐、席南华、苏育才、芮和兵

最后我必须再重复一下: 我国年青代数学家在模表示论和线性代数群上的工作, 远不止上面作为例子提到的这几项, 是值得大书特书的; 我国年青代数学家在代数 K 论、模论、同调代数、Hopf 代数、代数表示论、无限维 Lie 代数、超 Lie 代数和交换代数等分支上的工作是众多的, 也值得大写特写. 这都是我国代数学的重要组成部分, 限于时间和我的知识面, 只好在本文中从略. 我希望本文能起到抛砖引玉的作用, 更希望这些方面的专家能自己撰写这些方面的综述论文, 把它们综合起来就会成为全面的中国的代数学.

编者按: 本文是作者根据在中国数学科学与数学教育发展论坛 (2006 年 6 月 30 日 – 7 月 2 日, 杭州) 上的报告所写成的.

忆华罗庚老师 1950 年回到清华园执教[①]

万哲先

新中国成立不到半年, 1950 年 3 月 16 日华罗庚老师怀着满腔爱国热情, 舍弃了美国大学里活跃的科研环境和优越的生活待遇, 毅然回国, 回到清华园, 在清华大学执教. 华老师这一爱国行动受到党和政府的高度赞扬和清华大学师生的热烈欢迎. 在学期开始的时候, 清华大学算学系的课程表里就已宣告, 这学期有 "初等数论" 和 "广义矩阵论" 两门课程, 授课教师是华罗庚. "初等数论" 这门课是为算学系一年级学生开的, 选课的同学有王萼芳、张鸣华、解基培等人, "广义矩阵论" 这门课是为算学系高年级学生开的, 选课的同学有丁石孙、曾肯成等人. 华老师回到清华园后, 很快就走上了课堂. 他当时非常兴奋, 教了几年外国学生, 现在回来教自己的学生, 而自己学生的程度又让他赞不绝口. 选课的学生更是兴奋, 有幸有一位举世闻名的大数学家来给他们授课, 讲课又是非常清晰, 重点突出, 深入浅出, 引人入胜, 富有启发性. 我有幸被系里指定为 "广义矩阵论" 课程的助教, 从而有了直接向华老师学习并在他指导下做研究的机会. "初等数论" 课程的助教是迟宗陶学长, 他当时是研究生, 在闵嗣鹤教授指导下做硕士论文, 即将毕业.

在 "广义矩阵论" 这门课的第一节课上, 华老师列出了这门课的内容:

矩阵群 主要是典型群, 包括线性群、辛群、正交群和酉群, 讨论的问题是它们的结构、自同构和表示论等.

矩阵环 主要是各式各样的矩阵单环、如全矩阵结合单环、各种矩阵组成的单 Lie 环、单 Jordan 环等. 讨论的问题是它们的分类、自同构和表示.

矩阵几何 包括线几何、圆几何、Grassmann 几何、长方阵几何、对称

[①] 原文刊登于《传奇数学家华罗庚 —— 纪念华罗庚诞辰 100 周年》, 数学与人文, 第二辑, 北京: 高等教育出版社, 2010, 36–39.

矩阵几何、斜对称矩阵几何, 等等. 讨论的问题是: von Staudt 定理在各类矩阵空间上的推广, 及其对代数、函数论的应用, 它们的子几何等等.

矩阵域　典型域及它们上面的函数论.

华老师特别强调, 整个讲课计划完成以后, 将会给许多数学分支以崭新面貌. 华老师的这一讲课提纲把学生过去学过的及没有学过的许多知识都用矩阵这一工具串了起来, 并且指出了发展前途. 学生听了, 大开眼界, 极受启发, 获益良多.

第二节课开始讲第一章, 除环. 华老师认为除环这个抽象代数结构与成果丰富的域相距不远, 而除环本身又有许多问题不清楚, 值得探讨挖掘. 在这一章里他除了把域上一些熟知结果推广到除环上以外, 着重讲了他本人在除环上的三条重要定理.

定理 1　除环的半自同构必为自同构或反自同构.

这条定理是华老师 1949 年证明的, 后来 E. Artin 把它写进了他的 *Geometric Algebra* (1957 年出版) 一书中, 并把它称为华的漂亮定理.

定理 2　除环的真正规子除环必包在中心之中.

这个定理最初是由 H. Cartan 于 1947 年作为他所建立的除代数的 Galois 理论的应用而仅对除代数证明的. 随后, 1949 年 R. Brauer 和华老师对除环给出了不同的简单证明, 因此, 这个定理在文献中被称为 Cartan-Brauer- 华定理.

定理 3　不是域的除环的乘法群的导来链不终止于 $\{1\}$.

这个定理是 1950 年上半年他在讲授 "广义矩阵论" 课程期间证明的, 发表于 1951 年.

华罗庚老师在证明上述三个定理时, 着重运用了恒等式的技巧. 他在课堂里强调, 做代数就要会运用恒等式的技巧, 做分析就要掌握处理不等式的技巧.

第二章 "一维射影几何及二维线性群" 是 "广义矩阵论" 这个课程关键性的一章. 复数域上一维射影几何的基本定理通常被称为 von Staudt 定理, 是一条熟知的著名定理, 在有的复变函数论书上可以查到. von Staudt 之后, 许多数学家将它推广到各式各样的域上, 到 20 世纪 40 年代又有数学家试图推广到除代数上, 但他们遇到了除代数的半自同构是否都是自同构或反自同构这一难题, 解决不了. 华老师 1949 年不但对除代数, 而且对除环证明了这一难题, 即证明前述定理1, 从而除环上一维射影几何的基本定理就迎刃而解.

根据 von Staudt 定理, 一维射影空间中将调和点列变到调和点列的变换组成的群是二维广义射影变换群. 华老师在证明了 von Staudt 定理之后, 接着就自然过渡到讲述二级线性群, 它们的结构和自同构. "广义矩阵论" 这门课程讲完这些内容, 一学期也就结束了.

这学期华老师除了开了两门课并带领学生做研究以外, 还组织了算学系的综合讨论班, 请校内外的教师演讲, 介绍他们自己的新成果. 还参加高年级同学组织的 "婴儿" 讨论班, 对学生鼓励并指导, 还在算学系布告牌上出算学题征解, 等等. 他的办公室的门整天开着, 欢迎师生来找他讨论算学问题. 清华算学系的学术气氛一下子就活跃了许多.

1950 年下半年, 虽然华老师已被指定筹建中国科学院数学研究所, 1950—1951 学年度他仍然在清华大学继续讲授 "广义矩阵论" 这门课, 原先选课的学生丁石孙、曾肯成已经毕业, 留校任助教, 仍旁听这门课, 又有高年级学生肖树铁、殷涌泉等选修这门课. 这一学年他一共讲了四章: 第三章, 向量空间、矩阵和行列式, 第四章, 射影几何与仿射几何, 第五章, 长方阵几何, 第六章, 线性群的构造及自同构.

华罗庚老师在课堂上指出, 研究典型群的自同构有两种方法, 一种是 J. Dieudonné 采用的方法. 他的方法对于大的级数的典型群处理起来很顺利, 但对于小的级数的典型群显得很笨拙, 不好处理. 另一种方法是华老师创用的方法. 以 1948 年他确定特征 $\neq 2$ 的任意域 F 上辛群 $Sp_{2\nu}(F)$ ($\nu \geqslant 1$) 的自同构为例: 他从最小的 $\nu = 1$ 出发, 先确定 $Sp_2(F)$ 的自同构 (纠正了 Schreier 和 v.d.Waerdan 论文中的错误), 然后对 $\nu \geqslant 2$ 的情形用归纳法来确定 $Sp_{2\nu}(F)$ 的自同构. 这一方法也可以用于其他类型的典型群. 华老师在广义矩阵论课里就用他的方法来讲授线性群的自同构. 在第二章里他先确定当时可以确定其自同构的那些二维线性群的自同构, 在第六章里又用归纳法来确定级数 $\geqslant 3$ 的线性群的自同构. 这样他就把他的方法教给学生, 并带着学生一起来确定 J. Dieudonné 所不能解决几个级数较低的线性群的自同构. 这样学生既学了知识又学了做研究, 收获很大. 华老师和他学生关于典型群方面的工作被国外专家称为典型群中国学派的工作, 1975 年访华的美国数学家代表团将中国典型群方面的工作列为中国数学五项重要成就之一. 他对学生要求非常严格, 对研究工作要求很高, 强调要选有意义的问题做, 要有新的想法, 要创造不要依样画葫芦. 一旦学生有了新的想法, 取得点滴成果时, 他就鼓励. 另一方面他经常告诫学生, 不要眼高手低, 只要练好扎实的基本功, 做到 "拳不离手, 曲不离口", 踏踏实实地工作, 收获就会到来.

1951 年秋到 1952 年初和 1956 年秋到 1957 年春, 他又在中国科学院数学研究所主持典型群讨论班, 讲授他在清华大学 "广义矩阵论" 的讲义, 后来又指示我把工作继续做下去, 我就根据前六章的精神又续写了六章. 由于这十二章的大部分讨论的是典型群, 1963 年就以 "典型群" 为书名出版.

矩阵几何是华老师在昆明西南联大任教期间, 20 世纪 40 年代初开创的一个研究领域, 他先是研究复数域上的矩阵几何, 主要研究的问题是如何运用空间的不变量来刻画运动群. 1945–1946 学年度的下学期, 他曾在西南联大开设 "矩阵几何" 选课, 当时我是二年级学生, 因为他的授课时间与我选的必修课的上课时间冲突, 我没有听成他开的课, 失去了更早地跟他学习的机会. 1948 年, 他将对称矩阵几何扩充到特征 $\neq 2$ 的任意域上, 并指出仅用粘切这一不变量就可以刻画对称矩阵空间的变换群, 这一结果被他称为对称矩阵几何的基本定理. 接着他又将长方阵几何推广到任意除环上, 仍然是用粘切这一不变量就可以刻画运动群. "广义矩阵论" 的第五章是讲授长方阵几何学和它的一些代数和几何的应用. 这些内容于 1951 年发表. 他证明长方阵几何基本定理的方法与他证明对称矩阵几何基本定理的方法完全不同, 前者的关键是引进极大集这一概念, 使基本定理证明起来非常简捷. 所谓极大集就是 20 世纪 70 年代在图论中出现的极大完全子图这一重要概念, 而后者比华老师的极大集晚出现了二十年. 他在课上建议学生考虑用极大集的方法来处理对称矩阵几何, 这个问题我考虑很久, 直到 1993 年我引进了秩 2 极大集的概念, 利用它才重新证明了特征 $\neq 2$ 的任意域上对称矩阵几何的基本定理. 接着特征 $= 2$ 的任意域上对称矩阵几何和埃尔米特矩阵几何的基本定理, 也就迎刃而解, 这样我写了 *Geometry of Matrices* (1997 年出版) 这本书来怀念华罗庚老师.